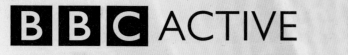

Maths

Higher Level

Graham Lawlor

Consultant: Rob Kearsley Bullen

Published by BBC Active, an imprint of Educational Publishers LLP, a part of the Pearson Education Group.
Edinburgh Gate, Harlow, Essex CM20 2JE England

© Graham Lawlor/BBC Worldwide Ltd 2002

BBC logo © BBC 1996. BBC and BBC Active are trademarks of the British Broadcasting Corporation

First published 2002

Third impression 2006

ISBN 13: 978-0-563-50128-2
ISBN 10: 0-563-50128-6

Illustrations by Hardlines Ltd

Printed by Canale, Italy

The Publisher's policy is to use paper manufactured from sustainable forests.

Contents

Introduction   4

Number

Fractions and decimals   8
Working with indices   10
Numbers in standard form   12
Ratio, proportion and percentages   14
Surds   16

Algebra

Solving simple equations   18
Rearranging formulae   20
Inequalities   22
Lines and equations   24
Simultaneous equations and graphs   26
Simultaneous equations and algebra   28
Using brackets in algebra   30
Multiplying bracketed expressions   32
Regions   34
Sequences and formulae   36
More on formulae   38
Equations of proportionality   40
Quadratic functions   42
Cubic functions   44
Graphing reciprocal functions   46
Solving equations with graphs   48
Trial and improvement   50
More complicated indices   52
Indices and equations   54
Factorising quadratic expressions   56
The difference of two squares   58
Algebraic fractions   60
Quadratic equations 1   62
Quadratic equations 2   64
More on simultaneous equations   66

Shape, space and measures

Congruence and similarity 68
Trigonometry in any triangle 70
The sine rule 72
The cosine rule 74
Problems in three dimensions 76
Transformations 78
Enlargements 80
Functions 82
Translating graphs 84
Stretching graphs 86
Vectors 1 88
Vectors 2 90
Motion graphs 92
Circle geometry 94
Circles: arcs and sectors 96
Volume and surface area 98

Data handling

Cumulative frequency 100
Histograms 102
Dispersion 104
Probability: the OR rule 106
Probability: the AND rule 108

Exam questions and model answers 110

Topic checker 118

Complete the facts 124

Answers to practice questions 132

Answers to check questions 141

Glossary 144

Last-minute learner 147

About Bitesize

GCSE Bitesize is a revision service designed to help you achieve success at GCSE. There are books, television programmes and a website, each of which provides a separate resource designed to help you get the best results.

The television programmes are available on video through your school or you can find out transmission times by calling **08700 100 222**.

The website can be found at http://www.bbc.co.uk/schools/gcsebitesize/maths

About this book

This book is your all-in-one revision companion for GCSE.
It gives you the three things you need for successful revision:

1 every topic, clearly organised and clearly explained
2 the most important facts and ideas highlighted for quick checking
3 all the practice you need – in the check questions in the margins, in the practice section at the end of each topic, and in the exam questions section at the end of this book.

Each topic is organised in the same way:

- **The bare bones** – a summary of the main points, an introduction to the topic, and a good way to check what you know

- **Key facts** highlighted throughout

- **Check questions** in the margin at the end of each section of the topic - have you understood this bit?

- **Remember tips** in the margin – extra advice on this section of the topic

- **Exam tips** in red – specific things to bear in mind for the exam

- **Practice questions** at the end of each topic – a range of questions to check your understanding.

The extra sections at the back of this book will help you check your progress and be confident that you know your stuff:

- a selection of **Exam questions and model answers** explained to help you get full marks, plus extra questions for you to try for yourself, with answers

- a **Topic checker** with quick questions in all topic areas

- a **Complete the facts** section

- a **Last-minute learner** – the most important facts in just four pages.

About the GCSE maths exam

Your school or college is in communication with an examination board, such as AQA, EdExcel or OCR, and will enter you officially for the maths exam. If you want more information about what you need to know, the specification of your exam can be downloaded from their websites.

The exam usually consists of three parts:

- Paper 1 covers Number and Algebra, Shape, Space and Measures, and Data Handling. You are not allowed to use a calculator on this paper.
- Paper 2 covers the same topics as Paper 1. A calculator is essential on this paper.
- Coursework assignments covering Using and Applying Mathematics, or a third examination paper designed to test this area.

Preparing for the exam

Check your equipment about a week before the exam. Make sure you have a couple of decent pens, at least two sharpened pencils or a refill for your mechanical pencil, a set of mathematical instruments (ruler, protractor, compass), your calculator and a transparent container for all of it (a sandwich bag is OK).

Check the condition of your calculator's batteries. If in any doubt, replace them with a new set about two days before Paper 2, as you will need your calculator for this paper. Make sure it is working properly – that way, you've got time to go out and replace them again if they're duff!

Maths exams tend to be in the morning. This is good, because you are usually sharper then than in the afternoon. You can take advantage of this by getting up extra early on the day of the exam, if you want to, and doing a bit of cramming using the Last-minute learner on page 147 of this book. Have a good breakfast so you have plenty of energy, and make sure you bring a drink with you.

If the weather's really hot, exam halls can be very unpleasant to sit in. Dress appropriately: some schools insist on uniform, others don't, but you can usually compensate for the weather in some way!

In the exam room

When you are issued with your exam paper, you must not open it immediately. However, there are some details on the front cover that you can fill in (your name, centre number, etc.) before you start the exam itself. If you're not sure where to write these details, ask one of the invigilators (teachers supervising the exam).

You are expected to write your answers in the exam booklet provided. This means that co-ordinate grids, triangular dotted paper, etc. are all printed for you, so you don't need to use extra paper.

During the exam

When you open your exam paper, have a look through it (all the way through). Make sure that all the pages are there.

Note how many questions there are. That will give you an idea of how long you can afford to spend on each question.

How the exams are marked

Most questions have a combination of method marks and answer marks. It should be obvious what these mean:

- **method marks** are awarded for correct working or approach
- **answer marks** are awarded for the right thing appearing in the answer space.

When questions have more than one part, you often have to carry an answer through from one part to the next. If you make a mistake, you will be penalised for that, but only once. In most cases, the examiners have instructions to follow-through your mistake and see if you've worked correctly from that point on. More work for them, but fair on you!

Organising your revision

Studying maths is different to studying other subjects, because the only way to learn maths is to do maths. Simply reading through your notes is not enough in maths, you do have to put in the time and effort and actually tackle the questions.

The best way to study is to take an hour-long session, split it into three fifteen-minute sessions and have three five-minute breaks. Don't save all the breaks up to the end – you won't learn as effectively. You need to make sure that you split your sessions up in this way, it really does improve your learning. Taking breaks in this way allows your brain time to sort out the information.

Techniques to help you remember

There are things that you can do to enhance your memory as you revise.

- **Shapes** – Imagine that you had to remember five different types of numbers, for example: primes, squares, cubes, triangular numbers, and the Fibonacci sequence.

  One easy way to remember these five facts is to draw them around a pentagon:

You can then add diagrams onto the pentagon. The act of actually drawing the pentagon and labelling it makes it a worthwhile learning activity. It gives you a hook to remember. The fact that the pentagon has five sides means that you know you have to remember five facts. This can work in any subject.

Imagine you had five categories of information to memorise and each category had three smaller sets of facts. You could develop your shape memory-jogger to look like the one we've drawn on the left.

Here, the five facts can be listed around the pentagon and each subsection of three facts can be listed around each small triangle.

If some of the sub-sections have only one fact, instead of using a triangle, use a circle. The shape memory-jogger is fun, creative and uses both sides of your brain.

- **Chunking** – This means grouping facts together. For instance, one of the easiest ways to remember phone numbers is to group the digits in threes. So if you can group material, it will make it easier to remember.

- **Images** – Making images is creative, involving you more actively in the process of revising, helping you to remember. In your revision notes, make pictures and images as often as you can. Good images:
    - are colourful
    - are lively and dynamic
    - can make you laugh
    - often have exaggerated aspects to them.

Author's note

Being the author of this book simply means I am part of a team. Therefore, I need to put on record my thanks to Nick Jones and the team at the BBC for their faith in me, in particular Ursula Faulkner, Rob Kearsley Bullen, Dr Simon Harden and also my colleague Jane Furlong of Soar Valley College in Leicester for her invaluable advice.

Finally, my thanks to my partner Judith Lawlor for her support during the writing of this book, without which I could never have completed the task.

THE BARE BONES

➤ All fractions can be represented as decimals.
➤ You can convert any fraction to a decimal by treating it as a division.
➤ Certain decimals can be changed to fractions by writing as whole numbers over a suitable denominator.

A Changing fractions to decimals

1 To convert a fraction into its equivalent decimal, treat it as a division. Check these examples using your calculator:

$\frac{3}{10} = 3 \div 10 = 0.3$

$\frac{5}{8} = 5 \div 8 = 0.625$

$\frac{1}{3} = 1 \div 3 = 0.33333333333...$- this is written $0.\dot{3}$

$\frac{4}{7} = 4 \div 7 = 0.5714285714...$- this is written $0.\dot{5}7142\dot{8}$

> **If a single digit repeats, a dot is placed over it. Two dots show the first and last digits in a repeating group.**

KEY FACT

2 Some of these decimals **terminate** (have a finite number of decimal places, then stop). These have denominators of the form $2^p \times 5^q$, where p and q are integers, e.g. $400 = 2^4 \times 5^2$. All other denominators, having prime factors other than 2 and 5, can form **recurring decimals**. These go on forever, but have a repeating pattern in their digits.

3 Note that the correct choice of numerator can cause denominators that normally produce recurring decimals to produce some terminating ones, e.g. $\frac{7}{14}$, $\frac{3}{12}$, etc.

Q Can you make a list of ten different denominators that produce terminating decimals?

B Changing terminating decimals to fractions

1 Changing a terminating decimal to a fraction is quite straightforward.
- Look at the number of digits (d) after the decimal point.
- Write these digits as a whole number (this will be the numerator of the fraction).
- Form the power 10^d. This is the denominator.
- Cancel the fraction to lowest terms.

2 Change 0.675 to a fraction.
There are 3 digits after the decimal point. The numerator is 675.
The denominator is $10^3 = 1000$.
The fraction is $\frac{675}{1000}$, which cancels to $\frac{27}{40}$.

3 Change 0.0042 to a fraction.
There are 4 digits after the decimal point. The numerator is 42.
The denominator is $10^4 = 10000$.
The fraction is $\frac{42}{10000}$ which cancels to $\frac{21}{5000}$.

Q Why wasn't the numerator written as 0042 in the example?

c Changing recurring decimals to fractions

1 In this section, r stands for a **recurring decimal**.
To change a recurring decimal to its equivalent fraction, follow these steps:
- Look at the number of digits in the repeating part of the decimal. This is called the **period** of the decimal. Use p to stand for this number.
- Multiply the decimal by 10^p.
- Subtract the original decimal from this new one. You should obtain an integer or a terminating decimal.
- This integer is a multiple of the required fraction. Which multiple it is determines the denominator.
- Form the fraction. Multiply top and bottom by 10 until the numerator is an integer.
- Cancel to lowest terms.

2 Change $0.\dot{6}$ to a fraction.
There is 1 recurring digit, so $p = 1$ and you multiply by $10^1 = 10$.
$10r = 6.666666666...$
$r = 0.666666666...$

Subtracting gives
$9r = 6$
$r = \frac{6}{9} = \frac{2}{3}$ (lowest terms)

3 Change $0.3\dot{1}5\dot{1}$ to a fraction.
There are 3 recurring digits, so $p = 3$ and you multiply by $10^3 = 1000$.
$1000r = 315.1151151151...$
$r = 0.3151151151...$

Subtracting gives
$999r = 314.8$ (the multiple is the 999th, so 999 is the denominator)
$r = \frac{314.8}{999} = \frac{3148}{9990}$ (multiply by 10 so numerator is an integer)
$r = \frac{1574}{4995}$ (lowest terms)

PRACTICE

1 Which of these denominators always produce terminating decimals, and which can produce recurring ones? Do not use a calculator!

(a) $\frac{}{8}$ (b) $\frac{}{9}$ (c) $\frac{}{10}$ (d) $\frac{}{25}$

(e) $\frac{}{24}$ (f) $\frac{}{23}$ (g) $\frac{}{32}$ (h) $\frac{}{80}$

(i) $\frac{}{90}$ (j) $\frac{}{125}$

2 Change these fractions to decimals:

(a) $\frac{7}{10}$ (b) $\frac{21}{100}$ (c) $\frac{22}{25}$

(d) $\frac{5}{6}$ (e) $\frac{3}{11}$ (f) $\frac{7}{60}$

3 Change these decimals to fractions:

(a) 0.44 (b) 0.0095 (c) 0.0625
(d) $0.\dot{1}$ (e) $0.3\dot{2}\dot{1}$ (f) $0.2\dot{2}\dot{6}$

When looking for ways to cancel fractions, divide top and bottom by prime numbers only – this is more efficient than simply 'counting up'.

Working with indices

THE BARE BONES

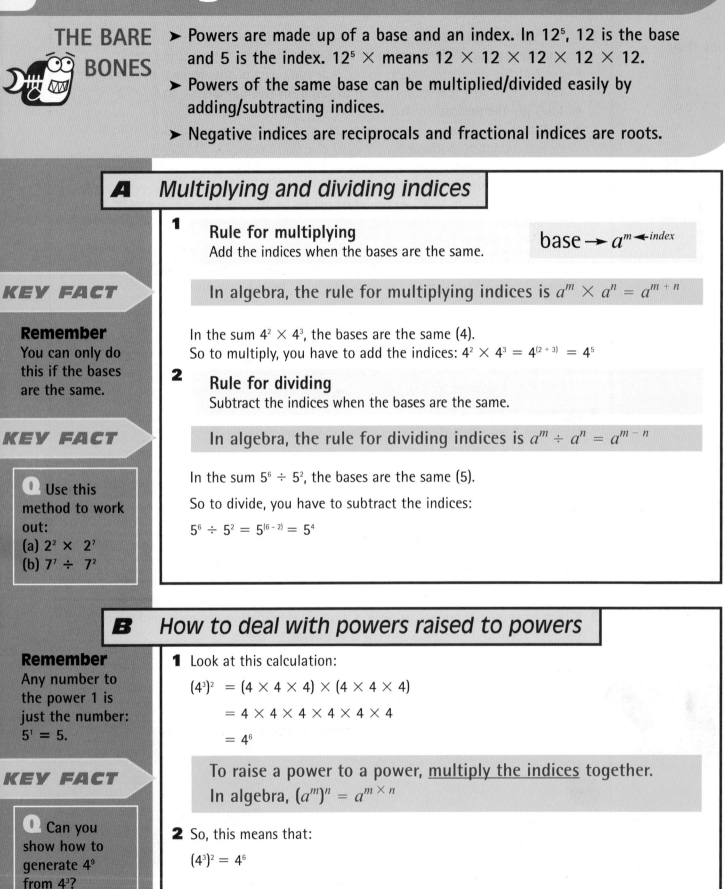

➤ Powers are made up of a base and an index. In 12^5, 12 is the base and 5 is the index. 12^5 × means $12 \times 12 \times 12 \times 12 \times 12$.

➤ Powers of the same base can be multiplied/divided easily by adding/subtracting indices.

➤ Negative indices are reciprocals and fractional indices are roots.

A Multiplying and dividing indices

1 Rule for multiplying
Add the indices when the bases are the same.

base → a^m ← index

KEY FACT

In algebra, the rule for multiplying indices is $a^m \times a^n = a^{m+n}$

Remember
You can only do this if the bases are the same.

In the sum $4^2 \times 4^3$, the bases are the same (4).
So to multiply, you have to add the indices: $4^2 \times 4^3 = 4^{(2+3)} = 4^5$

2 Rule for dividing
Subtract the indices when the bases are the same.

KEY FACT

In algebra, the rule for dividing indices is $a^m \div a^n = a^{m-n}$

In the sum $5^6 \div 5^2$, the bases are the same (5).

So to divide, you have to subtract the indices:

$5^6 \div 5^2 = 5^{(6-2)} = 5^4$

Q Use this method to work out:
(a) $2^2 \times 2^7$
(b) $7^7 \div 7^2$

B How to deal with powers raised to powers

Remember
Any number to the power 1 is just the number:
$5^1 = 5$.

1 Look at this calculation:

$(4^3)^2 = (4 \times 4 \times 4) \times (4 \times 4 \times 4)$

$= 4 \times 4 \times 4 \times 4 \times 4 \times 4$

$= 4^6$

KEY FACT

To raise a power to a power, <u>multiply the indices</u> together.
In algebra, $(a^m)^n = a^{m \times n}$

Q Can you show how to generate 4^9 from 4^3?

2 So, this means that:

$(4^3)^2 = 4^6$

C   How to deal with negative powers

1 When you use the index rule for division, you can end up with a **negative index**. For example, $3^2 \div 3^5 = 3^{2-5}$

$$= 3^{-3}$$

Remember
For this to be true, $a \neq 0$.

2 You can write it in a different way:

$$\frac{3^2}{3^5} = \frac{\cancel{3} \times \cancel{3}}{\cancel{3} \times \cancel{3} \times 3 \times 3 \times 3} = \frac{1}{3 \times 3 \times 3} = \frac{1}{3^3}$$

3 So $3^{-3} = \frac{1}{3^3} = \frac{1}{3 \times 3 \times 3} = \frac{1}{27}$

4 In general, $a^{-n} = \frac{1}{a^n}$

Q Can you write 2^{-5} as a fraction?

5 When two numbers are multiplied together and the answer is 1, the numbers are said to be **reciprocals** of each other. So the reciprocal of 4 is $\frac{1}{4}$ because when they are multiplied, the **answer is 1**.

D   How to deal with fractional indices

1 **Fractional indices** means **roots**.

2 You should know that $\sqrt{16} = 4$.

3 The rule of indices tells us that 16 is 16^1. If you use this rule, then $\sqrt{16} \times \sqrt{16} = 16$ (and 4×4 does equal 16).

Q Can you rewrite $x^{\frac{1}{4}}$?

4 $\sqrt{16}$ must be the same as $16^{\frac{1}{2}}$.

This makes sense because $16^{\frac{1}{2}} \times 16^{\frac{1}{2}} = 16^1 \left(\frac{1}{2} + \frac{1}{2} = 1\right)$.

5 The same argument applies to $\sqrt[3]{a}$. This can be written as $a^{\frac{1}{3}}$.

KEY FACT

In general: $\sqrt[n]{a} = a^{\frac{1}{n}}$

PRACTICE

1 Work out:

(a) $9^{\frac{1}{2}}$ (b) $9^{-\frac{1}{2}}$ (c) $9^{\frac{3}{2}}$

2 Work out the following, writing your answers in both index form and without using indices where possible.

(a) 14^0 (b) $(5^2)^{-3}$ (c) $36^{\frac{1}{2}}$ (d) $64^{-\frac{1}{2}}$

3 Work out $16^{\frac{1}{2}} \times 8^{\frac{1}{3}}$.

4 What is $49^{\frac{1}{2}} \times 49^{-\frac{1}{2}}$?

5 Calculate $64^{\frac{1}{2}} \div 2^3$.

Numbers in standard form

THE BARE BONES

➤ In science, you often need to use very large or very small numbers.

➤ 6.1×10^5, 9×10^{-3} and 8.01×10^{12} are examples of numbers in standard form.

➤ Numbers not in index form are in 'ordinary' form.

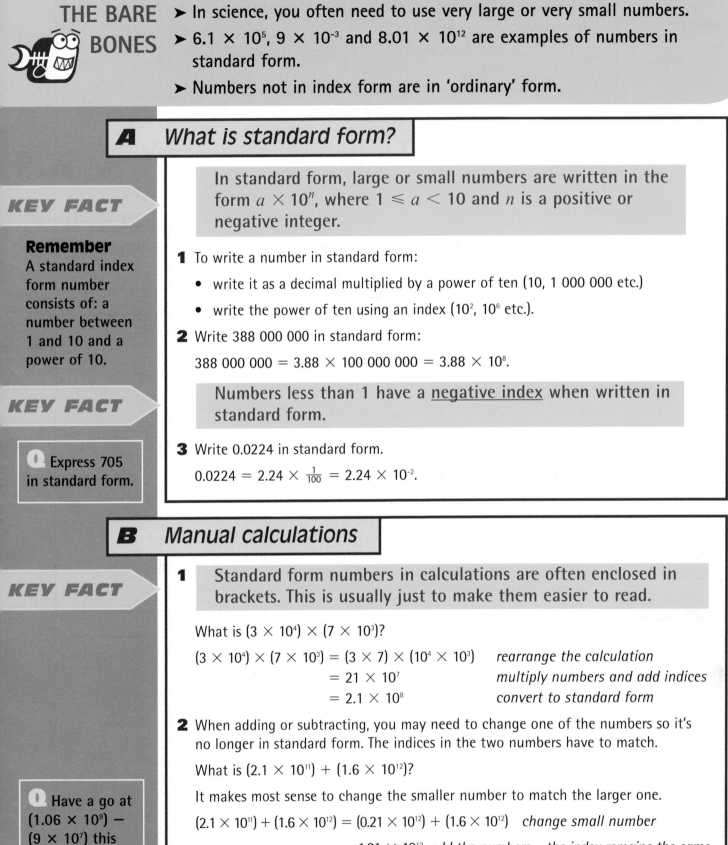

A What is standard form?

In standard form, large or small numbers are written in the form $a \times 10^n$, where $1 \leqslant a < 10$ and n is a positive or negative integer.

KEY FACT

Remember
A standard index form number consists of: a number between 1 and 10 and a power of 10.

1 To write a number in standard form:

• write it as a decimal multiplied by a power of ten (10, 1 000 000 etc.)

• write the power of ten using an index (10^2, 10^6 etc.).

2 Write 388 000 000 in standard form:

388 000 000 = $3.88 \times 100\ 000\ 000$ = 3.88×10^8.

Numbers less than 1 have a <u>negative index</u> when written in standard form.

KEY FACT

3 Write 0.0224 in standard form.

$0.0224 = 2.24 \times \frac{1}{100} = 2.24 \times 10^{-2}$.

Q Express 705 in standard form.

B Manual calculations

KEY FACT

1 Standard form numbers in calculations are often enclosed in brackets. This is usually just to make them easier to read.

What is $(3 \times 10^4) \times (7 \times 10^3)$?

$(3 \times 10^4) \times (7 \times 10^3) = (3 \times 7) \times (10^4 \times 10^3)$ *rearrange the calculation*

$= 21 \times 10^7$ *multiply numbers and add indices*

$= 2.1 \times 10^8$ *convert to standard form*

2 When adding or subtracting, you may need to change one of the numbers so it's no longer in standard form. The indices in the two numbers have to match.

What is $(2.1 \times 10^{11}) + (1.6 \times 10^{12})$?

It makes most sense to change the smaller number to match the larger one.

$(2.1 \times 10^{11}) + (1.6 \times 10^{12}) = (0.21 \times 10^{12}) + (1.6 \times 10^{12})$ *change small number*

$= 1.81 \times 10^{12}$ *add the numbers – the index remains the same*

Q Have a go at $(1.06 \times 10^9) - (9 \times 10^7)$ this way.

C On a calculator – the EXP button

1 On your calculator you have a button like this EXP , for entering numbers in standard form.

2 If you key in 4 . 5 EXP 7 , you should get `45000000` on your display. You have entered 4.5×10^7 into your calculator.

3 To key a number with a negative index, follow your calculator's instructions for negative numbers. On older calculators, to enter 7.3×10^{-8}, you might need to key in 7 . 3 EXP 8 +/−

or 7 . 3 EXP (−) 8

Q Find a calculation that causes a display error because the answer is too big.

D On a calculator – doing calculations

1 There is no difference between doing calculations with ordinary form numbers and standard form numbers. You can also mix them up in the same calculation. Just follow the usual rules.

KEY FACT

> When you use standard form numbers in a calculation, the answer may be within the calculator's normal display range, and be displayed without an index, so you may have to convert it yourself.

2 Calculate: $(8 \times 10^3) \times (4 \times 10^3)$. Key in:

8 EXP 3 × 4 EXP 3

to get 3.2×10^7.

Q What types of calculation don't need a second number?

3 Calculate: $(5 \times 10^8)^2$. Key in: 5 EXP 8 x^2 to get 2.5×10^{17}.

PRACTICE

1 Write in standard form:
(a) 600   (b) 2 000 000   (c) 41 000   (d) 955 000
(e) 0.0009   (f) 0.003   (g) 0.00000101   (h) 0.58

2 Write in ordinary form:
(a) 8×10^3 (b) 4×10^5 (c) 2.1×10^9 (d) 7.003×10^7
(e) 3×10^{-1} (f) 5×10^{-8} (g) 5.75×10^{-2} (h) 6.77×10^{-6}

3 Calculate 5^{75}. Give your answer in standard form, correct to 3 s.f.

4 If $x = 2 \times 10^{12}$ and $y = 3 \times 10^5$, find: (a) xy (b) $\frac{y}{x}$

5 Work out: (a) $(3.1 \times 10^{-4}) \times (4.3 \times 10^{12})$ (b) $(3.7 \times 10^{-6})^2$

6 Given that $M = \sqrt{\frac{n}{p}}$, find M in standard form when $n = 3.6 \times 10^2$ and $p = 1.6 \times 10^4$

Ratio, proportion and percentages

THE BARE BONES

➤ Ratios give us a way of comparing quantities.
➤ Proportion shows us that as one quantity increases or decreases, a corresponding quantity does the same.
➤ Percentages are commonly used to determine increases and decreases.

A Ratio

KEY FACT

> The ratio of one quantity to another is written as $a:b$ or $\frac{a}{b}$.

1 Maps are drawn to scale. A common scale is 2 cm to 1 km. This means that 2 cm on the map stands for a real distance of 1 km. Another way of saying this is 1:50 000. This is a **ratio**.

> Be careful! The order in which the comparison is made has to be clear.
> $a:b$ is not equal to $b:a$.

2 The two quantities must be in the same units and the answer must be in the simplest form. **Simplest form means it cannot be cancelled further.**

KEY FACT

> <u>Unitary ratios</u> are ratios that can be written in the form <u>1:n or n:1</u> The number n is either a decimal or a whole number.

3 Express this ratio in the simplest form – a length of 36 m to 48 m.

First find the **highest common factor** of 36 and 48. This is 12. Now divide by it. This means 36:48 cancels to 3:4.

4 Increase £4.50 in the ratio 3:2. Call the new amount £x; then
£x : £4.50 = 3:2. Write this as a fraction: $\frac{x}{4.50} = \frac{3}{2}$.

$$2x = £13.50$$
$$x = £6.75$$

Q How many parts can a ratio have?

B Proportion

1 If measurements are in direct proportion, this means that as one increases, the other increases by the same percentage.

2 Money is changed from one currency to another using the method of proportion.

3 In January 2002, the euro became legal tender in many European countries.
- On the day of checking, a euro (€) was worth about 62p. How many euros is £30 worth?
  Divide 30 by 0.62 this gives € 48.39 (€48 and 39 cents).
- Change € 25 to pounds.
  Multiply the € 25 by 62p = £15.50.

Q Do an Internet search to find out what £50 is worth today in euros.

C Percentages

All percentages are fractions over 100.
So $10\% = \frac{10}{100} = \frac{1}{10}$ of a quantity.

1 Finding a percentage of a quantity

Divide the percentage by 100 and then multiply by the quantity.

Find 12% of £46.

$\frac{12}{100} \times 46 = 5.52 = £5.52$

2 Expressing one quantity as a percentage of another quantity

Calculate 17 000 out of 20 000 as a percentage.

$\frac{17\,000}{20\,000} \times 100\% = 85\%$

3 Increasing a quantity by a percentage

Increase £10 by 30%.

Here £10 is 100%, so you need 130%.

$\frac{130}{100} \times 10 = £13$

4 Reducing a quantity by a percentage

Reduce £10 by 40%.

Now you need 60% of the original amount.

You have already seen this in type 1:
$\frac{60}{100} \times 10 = £6$

These are the four types of percentage question you might be asked in an exam.

Q Any percentage can be written as a decimal, e.g. 12% = 0.12. How would you write the other percentages in this section on decimals?

1 Write these ratios as simply as possible:

(a) 8g : 1kg   (b) 125p : £10

2 Convert the following sterling amounts to euros (take € 1 as 62p):

(a) £35   (b) £142   (c) £1500

3 Over a period of 8 months a hive of bees increased in number by 25% and then by another 42%. Originally there were 2800 bees; how many are there now?

4 When an iron bar is heated, it increases in length by 0.3%. If the increase in length is 2 cm, what was the original length of the bar?

5 In the last three weeks of a sale, prices were reduced by 25% and then another 30%. What was the final sale price of a coat that originally cost £120?

Surds

THE BARE BONES

➤ When a number is written as the product of two equal factors, that factor is called the square root of the number e.g. $4 = 2 \times 2$ means that 2 is the square root of 4 or $\sqrt{4} = 2$

➤ Notice, −2 is also a square root of 4, because $-2 \times -2 = 4$

A Rational numbers and surds

1 Integers and fractions are called **rational** numbers. The square roots of some numbers are rational, e.g. $\sqrt{4} = 2$, $\sqrt{36} = 6$, $\sqrt{9} = 3$

KEY FACT

$\sqrt{}$ stands for the positive square root only.

2 This is **not** true of all square roots. Numbers like $\sqrt{2}$, $\sqrt{7}$ etc. can't be written as fractions, and can't be expressed exactly as a decimal, though you can approximate to as many decimal places as you like. Numbers like this are called **irrational** numbers.

3 The only way to give an exact answer containing these numbers is to leave them in square root form, e.g. $\sqrt{2}$. In this form, they are called **surds**.

Q Can you define a surd?

B Simplifying surds

1 $\sqrt{12}$ can be written as $\sqrt{(4 \times 3)}$, which is the same as $\sqrt{4} \times \sqrt{3}$, or $2\sqrt{3}$.

The 'trick' is to split the number inside the root into a product, one of whose parts is a perfect square.

In the same way, $\sqrt{18} = \sqrt{9 \times 2} = \sqrt{9} \times \sqrt{2} = 3\sqrt{2}$.

2 Fractional surds can sometimes be simplified using a similar technique:

$$\sqrt{\frac{3}{25}} = \frac{\sqrt{3}}{\sqrt{25}} = \frac{\sqrt{3}}{5}$$

In this example, an equivalent fraction is used to create a perfect square in the denominator: $\sqrt{\frac{1}{18}} = \sqrt{\frac{2}{36}} = \frac{\sqrt{2}}{\sqrt{36}} = \frac{\sqrt{2}}{6}$

Q Can you describe the most important step in simplifying a surd?

C Multiplying surds

1 To multiply expressions containing surds, follow the same rules as you would use to expand brackets in algebra.

Simplify $\sqrt{3}(2 + \sqrt{7})$.

$\sqrt{3}(2 + \sqrt{7}) = \sqrt{3} \times 2 + \sqrt{3} \times \sqrt{7}$

$\qquad = 2\sqrt{3} + \sqrt{21}$.

C

2 In the example above, there are no like terms to collect, but when the same surd occurs in each bracket, the expansion can be simplified.

Simplify $(3 + \sqrt{2})(5 + \sqrt{2})$.

$$(3 + \sqrt{2})(5 + \sqrt{2}) = 3(5 + \sqrt{2}) + \sqrt{2}\,(5 + \sqrt{2})$$
$$= 3 \times 5 + 3\sqrt{2} + 5\sqrt{2} + \sqrt{2} \times \sqrt{2}$$
$$= 15 + 8\sqrt{2} + 2$$
$$= 17 + 8\sqrt{2}$$

Q Can you explain why $3\sqrt{2} + 5\sqrt{2} = 8\sqrt{2}$?

> The surd question is likely to turn up on paper 1 – that means you won't be able to verify the answer with your calculator, so check it very carefully.

D *Rationalising a denominator*

1 Sometimes, when you calculate an answer, the surd appears on the bottom of a fraction. It is usual to rewrite the answer so that the surd appears on the top instead, with a whole number in the denominator. This is called **rationalising** the denominator.

Rationalise $\dfrac{3}{\sqrt{10}}$

$$\frac{3}{\sqrt{10}} = \frac{3}{\sqrt{10}} \times \frac{\sqrt{10}}{\sqrt{10}} = \frac{3 \times \sqrt{10}}{\sqrt{10} \times \sqrt{10}} = \frac{3\sqrt{10}}{10}$$

This technique involves multiplying top and bottom by the surd found on the bottom.

2 If the denominator contains a combination of integers and surds such as $3 + 2\sqrt{7}$, create a new term by reversing the sign of the surd (in this case, $3 - 2\sqrt{7}$), then multiply top and bottom by it.

Q Evaluate the original surd expressions and the rationalised ones. Convince yourself that they represent the same number.

Rationalise $\dfrac{2}{3+2\sqrt{7}}$

$$\frac{2}{3+2\sqrt{7}} = \frac{2(3-2\sqrt{7}\,)}{(3+2\sqrt{7})(3-2\sqrt{7}\,)} = \frac{6-4\sqrt{7}}{9-6\sqrt{7}+6\sqrt{7}-4\times7} = \frac{6-4\sqrt{7}}{9-28} = \frac{6-4\sqrt{7}}{-19} = \frac{4\sqrt{7}-6}{19}$$

PRACTICE

Simplify these surd expressions:

1 $\sqrt{32}$ **2** $\sqrt{300}$ **3** $\sqrt{245}$ **4** $\sqrt{\dfrac{3}{16}}$ **5** $\sqrt{\dfrac{8}{9}}$ **6** $\sqrt{\dfrac{1}{50}}$

7 $\sqrt{7}\,(3 + \sqrt{3})$ **8** $\sqrt{2}\,(1 - \sqrt{2})$ **9** $\sqrt{10}\,(4 + 3\sqrt{10})$

10 $(2 + \sqrt{6})(1 + \sqrt{6})$ **11** $(4 + \sqrt{3})(3 - 2\sqrt{3})$ **12** $(10 - 2\sqrt{5})(1 - 5\sqrt{5})$

13 $\dfrac{3}{\sqrt{11}}$ **14** $\dfrac{2}{5\sqrt{6}}$ **15** $\sqrt{2} \div \sqrt{6}$

16 $\dfrac{\sqrt{3}}{2-\sqrt{3}}$ **17** $\dfrac{1+\sqrt{5}}{1-\sqrt{5}}$

Remember
There are many types of roots – cube roots ($\sqrt[3]{\ }$), etc. The surd expressions you will be asked to simplify in the exam usually only use square roots.

THE BARE BONES

➤ To rearrange an equation, add or subtract, or multiply or divide the same quantity from both sides.

➤ Whatever you do to one side of the equation, you must do exactly the same to the other side.

➤ To 'undo' an operation, perform the opposite or inverse operation.

A Solving equations with an unknown on one side

KEY FACT

An equation is like a puzzle. Finding the correct value for the letter in an equation solves the puzzle. This value is the solution to the equation.

Remember
$3x$ means $3 \times x$, so to remove the 3 you need to divide by 3.

1 You can solve this equation: $3x + 4 = 19$

$3x + 4 - 4 = 19 - 4$ First take 4 from both sides.

$3x = 15$

$\frac{3x}{3} = \frac{15}{3}$ Now divide both sides by 3.

So $x = 5$

2 Solve $5x - 1 = 34$

$5x - 1 + 1 = 34 + 1$ Add 1 to both sides.

$5x = 35$

$\frac{5x}{5} = \frac{35}{5}$ Divide both sides by 5.

So $x = 7$

Q Can you define the coefficient of x?

Always work through your solution one step at a time. Trying to do too much at once can cause mistakes.

B Solving equations with unknowns on both sides

1 Look at this equation: $4x - 6 = 3x + 3$

Remove the x term from the right-hand side.

$4x - 6 - 3x = 3x + 3 - 3x$ Subtract $3x$ from both sides.

$x - 6 = 3$

Remove the number term from the left-hand side.

$x - 6 + 6 = 3 + 6$ Add 6 to both sides.

$x = 9$

Q Use this method to solve these equations:

(a) $2y + 5 = y + 9$

(b) $9k + 8 = 45k$

C Where the x term contains a fraction

1 There are two cases:
either the unknown is the denominator of the fraction, or the denominator.

$$\frac{x}{5} - 7 = 10$$

$$\frac{x}{5} = 10 + 7 \quad (+ \text{ 7 to both sides})$$

$$\frac{x}{5} = 17$$

$$x = 85 \qquad (\times \text{ both sides by 5})$$

$$\frac{2}{x} + 3 = 7$$

$$\frac{2}{x} = 7 - 3 \quad (- \text{ 3 from both sides})$$

$$\frac{2}{x} = 4$$

$$2 = 4x \qquad (\times \text{ both sides by } x)$$

$$0.5 = x \qquad (\div \text{ both sides by 4})$$

Q Can you explain how you got an answer of 85?

D Equations with a negative x term

1 This equation has a **negative unknown** on one side: $5 - 6x = 3$

$$5 - 6x + 6x = 3 + 6x \qquad \text{Add } 6x \text{ to both sides.}$$

$$5 = 3 + 6x$$

$$5 - 3 = 3 + 6x - 3 \qquad \text{Subtract 3 from both sides.}$$

$$2 = 6x$$

$$\frac{2}{6} = \frac{6x}{6}$$

$$\text{So } x = \frac{1}{3}$$

Remember
Check the answer by substituting into the original equation.

2 Solve $5x + 2 = 1 - x$

$$5x + 2 + x = 1 - x + x \qquad \text{Add } x \text{ to both sides.}$$

$$6x + 2 = 1$$

$$6x + 2 - 2 = 1 - 2 \qquad \text{Subtract 2 from both sides.}$$

$$6x = -1$$

$$\frac{6x}{6} = \frac{-1}{6}$$

$$x = \frac{-1}{6}$$

Q Use this method to solve these equations:

(a) $-9n = 12$

(b) $1 - 4r = -2r$

Align the equal signs in your working. It makes your answer look neater and it is easier to read your working.

PRACTICE

Solve these equations and check your answers:

1 $3x + 2 = 23$

2 $5y + 7 = 52$

3 $8d + 3 = 19$

4 $10f + 7 = 127$

5 $6c - 1 = 5c + 4$

6 $7h - 3 = 11h - 5$

7 $4x + 5 = 10x - 13$

8 $3x + 7 = 3 - x$

9 $9 - \frac{1}{3}k = 2k - 5$

10 $4 - 8p = 2 - 5p$

Rearranging formulae

THE BARE BONES
➤ Rearranging formulae means getting the subject on one side and everything else on the other side of the equals sign.
➤ It is also known as changing the subject, or transformation of formulae.

A Changing the subject – types 1 and 2

KEY FACT

> Changing the subject of a formula means rearranging a formula to get one letter on its own and all of the other letters on to the other side of the equals sign.

Remember
The plural of formula is 'formulae', but you might see it written as 'formulas' in the exam.

1 It is also known as **transformation** of formulae.

Type 1: When x is not 'bound' up with anything else:

$$x + a = b$$
$$x + a - a = b - a \quad \text{Here subtract } a \text{ from both sides.}$$
$$x = b - a$$

Type 2: When x is 'bound' in a multiplication:

$$xy = a$$
$$\frac{xy}{y} = \frac{a}{y} \qquad \text{Divide both sides by } y.$$
$$x = \frac{a}{y}$$

2 Look at this formula: $2x^2y = z$

$$\frac{2x^2y}{2y} = \frac{z}{2y} \qquad \text{Divide both sides by } 2y.$$
$$x^2 = \frac{z}{2y} \qquad \text{Now cancel the } 2y\text{s on the left.}$$
$$x = \sqrt{\frac{z}{2y}} \qquad \text{Remove the square by finding the square root of both sides.}$$

Q Make x the subject of this formula:

$x + y = p$

B Combinations of types 1 and 2

Remember
The equals signs should be directly under each other.

1 Make x the subject of this formula:

$$q + 6x = p$$
$$q + 6x - q = p - q \qquad \text{Isolate the } x\text{s first by taking } q \text{ from both sides.}$$
$$6x = p - q$$

2 You need only one x on the left-hand side, so 'undo' $6x$ by dividing each side by 6: $x = \frac{p - q}{6}$

Q How else could $\frac{p-q}{6}$ be written?

C What to do if the subject has a minus sign

1 Sometimes you may arrange a formula only to find that the left-hand side contains the subject with a minus sign in front of it, for example $-A = sx - 2t$.

KEY FACT

> If this happens, multiply both sides of the equation by –1.

This has the effect of changing the sign of **every term** in the formula, because: $-A \times -1 = A$.

If $-A = sx - 2t$, then $A = -sx + 2t$. You write this as $A = 2t - sx$.

2 Make r the subject of $3M - 2r = 4N$.

$$-2r = 4N - 3M \qquad \text{Subtract } 3M \text{ from both sides.}$$

$$-r = 2N - \tfrac{3M}{2} \qquad \text{Divide both sides by 2.}$$

$$r = -2N + \tfrac{3M}{2} \qquad \text{Multiply both sides by } -1.$$

$$r = \tfrac{3M}{2} - 2N \qquad \text{Rearrange right-hand side for neatness.}$$

Q What is the effect of multiplying a term by –1?

D When factorisation is required

Make sure you show your working. It is important to let the examiner see how you have worked out the answer.

1 If the letter you want to make the subject of a formula occurs in **two terms**, you may need to **collect** these terms and then **factorise** them.

2 Make x the subject of $2x = px + q$.

$$2x - px = q \qquad \text{Subtract } px \text{ from both sides.}$$

$$(2 - p)x = q \qquad \text{Factorise the left-hand side.}$$

$$x = \tfrac{q}{2 - p} \qquad \text{Divide both sides by } 2-p.$$

Q Make n the subject of:

$$V + sn = tn$$

PRACTICE

In these questions, make x the subject of the formula:

1 $x + 9 = r$

2 $x - z = a$

3 $5x + 4y = 16$

4 $mx^2 = c$

5 $\tfrac{1}{4}x = m$

6 $\tfrac{x}{p} = p + c$

7 $\tfrac{mx}{b} = c$

8 $5 - fx = 3x + p$

In these questions, make y the subject of the formula:

9 $2x + 5y = 9$

10 $x - 2y = 10$

Remember

In question 5, $\tfrac{1}{4}x$ is the same as $\tfrac{x}{4}$.

Inequalities

THE BARE BONES
➤ An inequality is a mathematical statement describing a range of values.
➤ Inequalities can be shown on a number line.
➤ To solve inequalities using algebra you apply the same techniques used in equations and rearranging formulae.

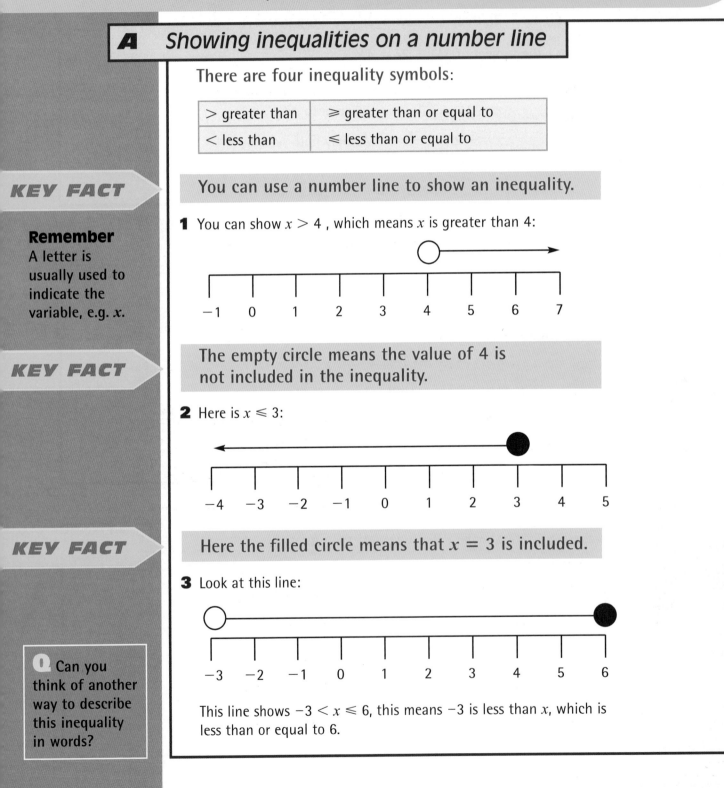

A **Showing inequalities on a number line**

There are four inequality symbols:

$>$ greater than	$\geq$ greater than or equal to
$<$ less than	$\leq$ less than or equal to

KEY FACT

Remember
A letter is usually used to indicate the variable, e.g. x.

You can use a number line to show an inequality.

1 You can show $x > 4$, which means x is greater than 4:

KEY FACT

The empty circle means the value of 4 is not included in the inequality.

2 Here is $x \leq 3$:

KEY FACT

Here the filled circle means that $x = 3$ is included.

3 Look at this line:

Q Can you think of another way to describe this inequality in words?

This line shows $-3 < x \leq 6$, this means -3 is less than x, which is less than or equal to 6.

B Solving inequalities

> You can solve equations to find an unknown number.
> You can solve inequalities to find a range of numbers.

KEY FACT

Remember
Do the same
thing to both
sides of an
inequality. If you
multiply or
divide by a
negative
number, the
direction of the
inequality is
reversed.

Q Can you
explain why
both sides were
divided by 4?

1 So $4 > -5$ is an inequality and it is a true statement.

But what happens if you:

(a) add 5 to both sides?   (c) subtract 3 from both sides?

(c) multiply both sides by 6?   (d) divide both sides by 2?

(e) multiply both sides by -4 (f) divide both sides by -2?

Investigate each of these statements and convince yourself.

You should find that when you multiply or divide by a negative quality, the sense of the inequality is **reversed**.

2 Work out this inequality:

$$4x + 5 > 29 \qquad \text{Subtract 5 from both sides.}$$

$$4x + 5 - 5 > 29 - 5$$

$$4x > 24 \qquad \text{Divide both sides by 4.}$$

$$\frac{4x}{4} > \frac{24}{4}$$

$$x > 6$$

3 This is a bit more complex:

$$-3 \leqslant 4x + 5 < 12$$

Treat this as two inequalities.

Step 1	Work out $-3 \leqslant 4x + 5$	the answer is $-2 \leqslant x$.
Step 2	Work out $4x + 5 < 12$	the answer is $x < \frac{7}{4}$.
Step 3	Put them together	the final answer is $-2 \leqslant x < \frac{7}{4}$.

PRACTICE

Solve these inequalities:

1 $2x + 7 \geqslant 3x + 2$ **2** $7x + 3 > 5x - 4$

3 $8 - 6a \leqslant 7$ **4** $1 - 6t \leqslant 9$

5 $-6d < 30$ **6** $-2w \geqslant 5$

7 $\frac{1}{4}f + 3 \geqslant 1$ **8** $6 > 2p + 3 > 4$

9 $1 \leqslant 5r + 2 \leqslant 12$ **10** $-3 < \frac{1}{3}t + 2 < 5$

Lines and equations

THE BARE BONES

➤ Every straight line on a co-ordinate grid has an equation satisfied by all the points on it.

➤ The gradient of a line is its slope or steepness, and is equal to the coefficient of x in the equation.

➤ All lines with the same gradient are parallel. The position of a particular line depends on the y-intercept, the value of y when $x = 0$.

A Graphs of straight lines

1 Every straight line on a co-ordinate grid can be written in the form $y = mx + c$.
So in the equation $y = 2x - 3$, $m = 2$ and $c = -3$.

> **KEY FACT**
>
> m is the number multiplying x (the coefficient of x), and represents the gradient or steepness of the line.
>
> c gives the position where the line crosses the y-axis and is called the y-intercept of the line.

B Calculating gradient

1 To calculate the gradient of a line on a diagram:
- select two points on it;
- work out the increase in x from one point to the other, and the increase in y;
- $\dfrac{\text{increase in } y}{\text{increase in } x}$ gives the gradient.

2 Find the gradient of the line in the diagram.

Q Why is the 'increase in y' negative?

The two chosen points are $A(4, 5)$ and $B(8, 3)$.
From A to B, the increase in x is 4 units, and the increase in y is –2 units.

The gradient is $m = \dfrac{\text{increase in } y}{\text{increase in } x} = \dfrac{-2}{4} = -\dfrac{1}{2}$

C Finding the equation of a line

1 Once the gradient of a line is known, its equation can be found by substituting x and y from a known point on the line. This will determine the value of c.

2 Find the equation of the line in the diagram above, and determine where it crosses the y-axis.

Q Sometimes you may not need to calculate c - why?

You know that $m = -\frac{1}{2}$. So the equation of the line is $y = -\frac{1}{2}x + c$.
Pick one of the points on the line, say (4, 5). Substitute for x and y:
$y = -\frac{1}{2}x + c$, so $5 = -\frac{1}{2} \times 4 + c$
$5 = -2 + c$, so $c = 7$.
So $y = -\frac{1}{2}x + 7$. This means the line crosses the y-axis at (0, 7).

D Equations of the form $ax + by = k$

1 Equations of this type also represent straight lines. Examples are $3x + y = -2$, $2x - 5y = 0$, etc.

To find the y-intercept, substitute $x = 0$ into the equation.

Find the y-intercept of the line with equation $3x - 2y = 2$.

Substitute $x = 0$: $3x - 2y = 2$
$$3 \times 0 - 2y = 2$$
$$-2y = 2, \text{ so } y = -1.$$
The line crosses the y-axis at $(0, -1)$.

2 To determine the gradient, make y the subject of the equation and use the coefficient of x.

Find the gradient of the line with equation $3x - 2y = 2$.

First, make y the subject: $y = \dfrac{3x - 2}{2} = \dfrac{3}{2}x - 1$

The gradient of the line is $\frac{3}{2}$ (or 1.5).

Note that the value of c can also be found from the rearranged equation.

Q Why are equations like this equivalent to $y = mx + c$?

E Perpendicular lines

Q What is the gradient of any line perpendicular to $y = \frac{1}{2}x$?

1 Two lines that intersect at right angles have the following property: the product of their gradients must equal -1.

2 The equation of the line perpendicular to $y = 3x - 5$, passing through the point $(2, 1)$, is $y = -\frac{1}{3}x + \frac{5}{3}$.

Any gradients that turn out to be fractions should be kept in fractional form. Avoid converting to a decimal unless it terminates.

PRACTICE

1 Write the gradient and y-intercept of lines with the equations:

(a) $y = 7x + 3$　　(b) $y = 3x - 5$　　(c) $y = x + 12$　　(d) $y = 3x - 2$

2 Find the gradient and y-intercept of the lines with the following equations:

(a) $3x + y = 4$　　(b) $2x - y = 7$　　(c) $x - 3y = 7$　　(d) $x + 5y = 9$

3 The gradient of a line is 4 and its y-intercept is at $(0, 5)$. What is the equation of the line?

4 A line passes through $(0, 6)$ and $(2, 8)$. Find the equation of the line.

5 Find the equation of the line that passes through the point $(0, 1)$ and is parallel to $y = 2x - 1$.

6 Find the equation of the line that is perpendicular to $y = 8 - 2x$, and intersects it at the point where $x = 3$.

THE BARE BONES

➤ A pair of equations, containing x and y, that are both true at the same time, are called simultaneous equations. There is usually a value of x and a value of y that satisfy both equations.

➤ If the equations are linear, they form straight lines. These usually cross. The coordinates of the point of intersection give the solution to the simultaneous equations.

A Simultaneous linear equations

1 The values $x = 3$ and $y = -1$ are a solution of the equation $y = x - 4$. They are also a solution of $2x + 3y = 3$. The graphs of these two equations cross at $(3, -1)$.

The solution of the **simultaneous** equations $y = x - 4$ and $2x + 3y = 3$ is $x = 3$, $y = -1$.

Q When would two graphs <u>not</u> cross each other?

B Solving simultaneous equations

1 To solve a pair of simultaneous equations, plot their graphs on the same axes. The solution is given by the co-ordinates of the point where the lines intersect.

2 Solve the simultaneous equations $y = 4 - x$ and $6y + 2x = 12$. Both the numbers in the solution are positive.

It's easy to draw the graph of $y = 4 - x$, but $6y + 2x = 12$ needs to be rearranged:

$6y = 12 - 2x$, so $y = 2 - \frac{1}{3}x$.

Now plot the graphs, using suitable co-ordinates:

The lines cross at $(3, 1)$, so the solution to the equations is $x = 3$ and $y = 1$.

Q Check that these numbers satisfy both the original equations.

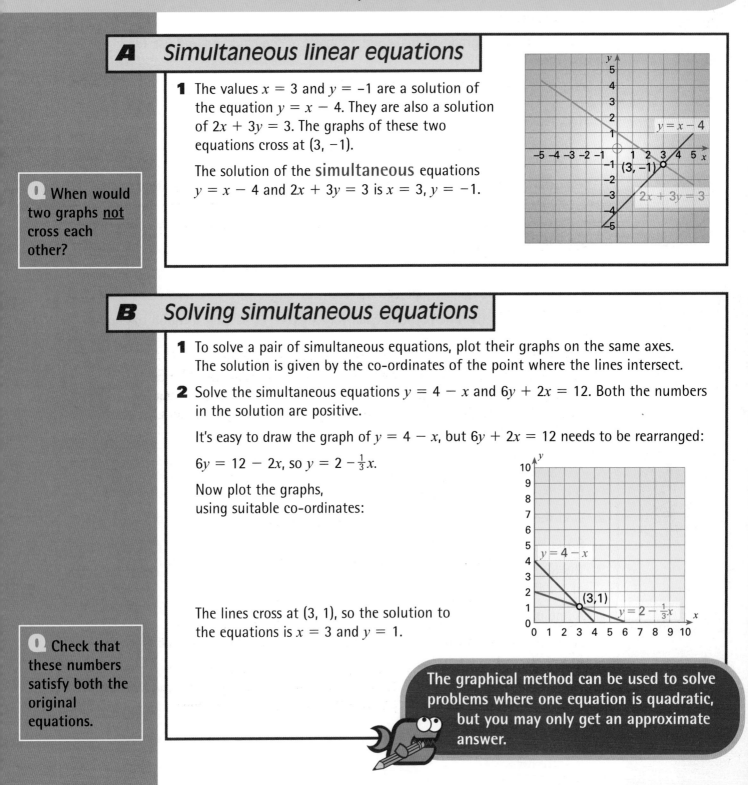

The graphical method can be used to solve problems where one equation is quadratic, but you may only get an approximate answer.

c   Solving problems using simultaneous equations

1 Some problems can be solved using simultaneous equations. The solution involves writing out the problem in algebraic form first, then solving the equations in the usual way. A problem is that you may not know the range in which the solution lies, so you may need more than one attempt to draw the graph.

2 Consider this problem: 3 apples and 2 bananas cost 81p. 2 apples and 5 bananas cost £1.20. How much does one of each fruit cost?

First choose letters to stand for the unknown quantities: say, x pence for an apple and y pence for a banana.

Then $3x + 2y = 81$ and $2x + 5y = 120$.

Rearranging, the two equations become
$$y = \frac{81 - 3x}{2} \text{ and } y = \frac{120 - 2x}{5}$$

To choose the scale for the graph axes, use common sense. Neither fruit is likely to cost more than about 30p, so use a scale from 0 to 30 on both axes. Plot the graphs.

The graphs give the solution $x = 15$, $y = 18$, so the answer is that an apple costs 15p and a banana 18p.

Remember

Straight lines have equations of the form $y = mx + c$ or $ax + by = k$.

Q Does it matter which you choose to be x and which you choose to be y?

PRACTICE

1 Draw the graphs of $x + y = 10$ and $2x + y = 13$. Hence solve the equations simultaneously.

2 Call the cost of hiring a car £y. Graylaw motors hires cars for a fee of £$(15 + 5x)$, where x is the number of days you hire the car for. Judelaw cars hire cars for £$(7.5x)$ per day. Write two equations to show the cost of hiring from each company. Draw graphs of each of the equations. Use your graphs to find the number of days for which the cost of hiring is the same for both companies, and how much this costs.

3 A business wants to have some booklets printed. Jayprintwell charges £y cost to print x booklets by using the equation $y = 1 + 0.5x$. Chargelightly printers say they will charge £15 plus 30p per booklet.

(a) Write the equation describing Chargelightly's pricing.
(b) Draw a graph to find:
   (i) the number of booklets that both companies would charge the same to produce;
   (ii) the cost of printing that number of booklets.

4 The points with co-ordinates (2, 2) and (5, 11) lie on the line with the equation $ax - by = 4$. Use simultaneous equations to find the values of a and b.

Remember to substitute the given co-ordinates into the equation of the line. Remember that the lines you use to solve the simultaneous equations aren't anything to do with the line $ax - by = 4$: it's as if they're in a 'different universe'.

THE BARE BONES

➤ By combining equations, it's possible to make one of the unknown letters disappear, leaving you with a simple equation in x or y to solve. This is called elimination.

A The Elimination method

1 Sometimes it's possible just to add two equations together and perform an elimination. It's useful to label equations with letters or numbers as you work. Consider these simultaneous equations:

$x - y = 3$ (i)
$x + y = 11$ (ii)

Adding them together gives
$x - y + x + y = 3 + 11$

Simplifying, $2x = 14$.   (iii)
The y's have been eliminated. Solving this equation, $x = 7$.

Substitute for x into (i), to find the value of y
$7 - y = 3$, so $y = 4$
Check: substitute x and y into equation (ii): $7 + 4 = 11$ ✓

> **KEY FACT**
>
> Always use one equation for the calculation and the other equation for the check.

2 The same result can be obtained by subtracting equation (i) from equation (ii).

$x + y - (x - y) = 11 - 3$
$x + y - x + y = 8$
$2y = 8$ (iv)

The x's have been eliminated.
Solving this gives $y = 4$, as before. $x = 7$ follows by substitution.

Q Solve
$x + 2y = 6$
$x - 2y = 10$
this way.

B Multiplying an equation

1 You can only eliminate one of the letters by adding or subtracting equations if the coefficients match. Sometimes, you need to 'force' a match by multiplying one of the equations first.

2 Solve $2m + 3n = 27$ (i)
                    $m + n = 12$ (ii)

Multiply equation (ii) by 2, to match the coefficients of m:
                    $2m + 2n = 24$ (iii)

Now subtract equation (iii) from equation (i):
$2m + 3n - (2m + 2n) = 27 - 24$, so $n = 3$

Q Multiply one of the equations to eliminate n instead of m. Check that you get the same solution.

B

Substitute for n into equation (i):
$$2m + 9 = 27$$
$$2m = 18, \text{ so } m = 9$$
Check: Substitute m and n into equation (ii) $9 + 3 = 12$ ✓

Never forget to carry out the final check, because it will always show up an error if you have made one.

C Equations where both coefficients need to be multiplied

Remember
f you have a graphical calculator, you can do a quick graph check of your solution.

1 Sometimes just multiplying one of the equations isn't enough to get a match. You need to multiply *both* equations. This can occasionally mean that quite large numbers appear in the new equations.

2 Solve $3x + 2y = 46$ (i)
         $2x - 5y = 18$ (ii)

To eliminate x, multiply equation (i) by 2 and equation (ii) by 3:
     $6x + 4y = 92$ (iii)
     $6x - 15y = 54$ (iv)

Equation (iii) − equation (iv):
$19y = 38$, so $y = 2$
Substitute y into equation (i):
     $3x + 4 = 46$
   $3x = 42$, so $x = 14$
Check: substitute for x and y in equation (ii): $28 - 10 = 18$ ✓

Q What would you multiply the equations by to eliminate y?

1 Solve each pair of simultaneous equations:

(a) $x + y = 7$
    $x - y = 3$

(b) $a + b = 11$
    $a - b = 3$

(c) $2m + n = 19$
    $m - n = 8$

(d) $3t + r = 25$
    $t - r = -1$

(e) $3x + 2y = 23$
    $2x + y = 14$

(f) $4c + 2d = 40$
    $3c + 3d = 48$

(g) $2e + f = 31$
    $3e + 2f = 50$

(h) $2p + 2q = 26$
    $3p + q = 43$

(i) $2a + b = 5$
    $a + 3b = 5$

(j) $p + 2q = 8$
    $2p + 3q = 14$

Remember
Solve the equations by eliminating v.

2 The cost of a car service, c pounds, for n hours of servicing is found by using the formula $c = a + bn$. It costs £100 for a 5-hour service and £76 for a 2-hour service. Find the values of a and b.

3 The points with co-ordinates $(1, 3)$ and $(3, \frac{9}{2})$ lie on the line with the equation $ax - by = 2$.

(a) Find the values of a and b.

(b) Find the gradient of the line.

If a variable has a positive coefficient in one equation and a negative coefficient in the other, choose to eliminate the negative one. You're less likely to make a mistake <u>adding</u> equations.

THE BARE BONES

➤ Expanding brackets means multiplying terms to remove brackets from equations or expressions.

➤ To expand an expression, you need to multiply each term inside the bracket by the term outside the bracket,
e.g. $3(x + y) = 3 \times x + 3 \times y = 3x + 3y$.

A Expanding brackets

KEY FACT

> To expand brackets, multiply everything in the bracket by the term outside the bracket.

1 Expand $3(2a + b) = 6a + 3b$.

2 Expand $6(y + 3) = 6y + 18$.

Remember
$4x^1 \times x^1$

$= 4x^1 \times x^1$
$= 4x^{1+1}$
$= 4x^2$

It is a common mistake to forget to multiply the **second** term.
If you worked out $6y + 3$ for the answer to the second equation, you'd be wrong!

3 Expand $3(2x + 4) = 6x + 12$.

This example uses the fact that $3 \times 2x = 6x$.

4 Expand $x(4x + 9) = 4x^2 + 9x$.

Notice here that the x is the term outside the bracket and that $x \times x = x^2$.

5 Expand $5x(3x - 2) = 15x^2 - 10x$.

Notice that $5x \times 3x = 5 \times 3 \times x \times x = 15x^2$.

Q What would $5x \times 3x^2$ expand to?

B Factorising expressions

KEY FACT

> Factorising is the opposite of expanding brackets.
> Find the highest common factor (HCF) of all of the terms in the expression you are trying to factorise.
> The HCF must appear outside the brackets.

Factorise $4x^2 + 8x$.

1 Here the HCF is the largest term that is a factor of $4x^2$ and $8x$. The HCF must be $4x$.

2 So you now have $4x^2 + 8x = 4x(? + ?)$

3 Ask yourself, 'What do I multiply $4x$ by, to make $4x^2$, and what do I multiply $4x$ by to make $8x$?'

4 So the final answer $= 4x(x + 2)$.

Q What is the HCF of $15x^2y$ and $20xy^2$?

C Solving equations with brackets

Method 1

1 Solve $4(x + 3) = 28$

2 Expand the brackets:

$$4x + 12 = 28$$

3 Solve it like any other equation:

$$4x + 12 = 28$$
$$4x = 16$$
$$x = 4$$

Method 2

1 Solve $4(x + 3) = 28$

2 $4(x + 3)$ is a product, it is $4 \times (x + 3)$

3 So divide by 4 to undo the multiplication:

$$\frac{4(x + 3)}{4} = \frac{28}{4}$$
$$x + 3 = 7$$
$$x = 4$$

Q Try both methods to solve this equation:

$4(x + 3) = 36$

D Solving equations with two sets of brackets

Solve $3(x + 4) + 5(x - 6) = 5x - 3$.

1 Expand the brackets and simplify:

$$3x + 12 + 5x - 30 = 5x - 3$$
$$8x - 18 = 5x - 3$$

2 Solve the equation. You should find that:

$$x = 5$$

Q Can you explain where the '$- 30$' came from in?

PRACTICE

Solve these equations:

1 $2(x + 5) = 18$

2 $5(x - 2) = 40$

3 $4(3x - 7) - 5(2x - 4) = 3x$

4 $2(7x - 7) - 2(4x + 5) = 21 - 4x$

5 $5 + (7x - 7) - 2(4x + 5) = 2x + 4$

6 $6(4x - 7) + 3(13 - 3x) = 20x - 23$

7 $8(2x - 9) - 2(9 - 2x) = 14x + 3$

8 Factorise:

    (a) $14x^2 + 7x$ **(b)** $36y^2 - 9y$

9 Factorise:

    (a) $15y^4 + 25y^2$ **(b)** $100a^2 + 20ab^3$

THE BARE BONES

➤ You need to know how to expand expressions, such as $(w + x)(y + z)$.

➤ Both terms in the first brackets have to be multiplied by both terms in the second, resulting in four new terms.

A Bracketed expressions

1 Look at this rectangle:

Q Can you use this idea to expand $(2w + x)(y + z)$?

2 The area of this rectangle is $(w + x)(y + z)$.

3 You can write this as the combined area of the four smaller rectangles:

$wy + wz + xy + xz$

B More difficult expansions

Remember
Each term in the first bracket is used to multiply everything in the second bracket. So $(x + 7)(x - 2)$ can be written as $x(x - 2) + 7(x - 2)$.

1 Expand and simplify:

$(x + 7)(x + 2)$

$= x(x + 2) + 7(x + 2)$

$= x^2 + 2x + 7x + 14$

$= x^2 + 9x + 14$

2 Expand and simplify:

$(x + 5)(x - 2) = x(x - 2) + 5(x - 2)$

Here you take every term in the first bracket and multiply the second bracket by it, in turn.

$= x^2 - 2x + 5x - 10$

$= x^2 + 3x - 10$

Q Can you explain where the $-x^2$ came from?

3 Expand and simplify:

$(x + 7)(3 - x) = x(3 - x) + 7(3 - x)$

$= 3x - x^2 + 21 - 7x$

$= 21 - 4x - x^2$

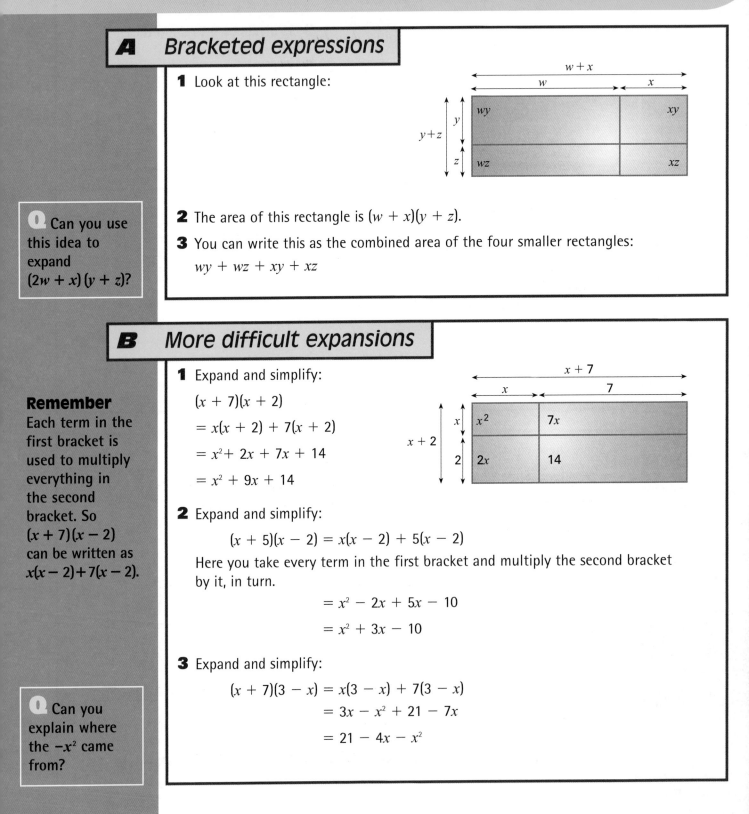

C Expanding a squared expression

1 Expand and simplify:

$$(x + 4)^2 = (x + 4)(x + 4)$$
$$= x(x + 4) + 4(x + 4)$$
$$= x^2 + 4x + 4x + 16$$
$$= x^2 + 8x + 16$$

2 Expand and simplify:

$$(x - 9)^2 = x(x - 9) - 9(x - 9)$$
$$= x^2 - 9x - 9x + 81$$
$$= x^2 - 18x + 81$$

Q Could you now expand $(x + 5)^2$?

KEY FACT

In general $(x + a)^2 = x^2 + 2ax + a^2$ and
$(x - a)^2 = x^2 - 2ax + a^2$

D A special expansion

1 Expand and simplify:
$$(x + 9)(x - 9) = x(x - 9) + 9(x - 9)$$
$$= x^2 - 9x + 9x - 81$$
$$= x^2 - 81$$

2 There is a special expansion that you must learn.
It is called the **difference of two squares**.

KEY FACT

In general $(x + a)(x - a) = x^2 - a^2$

You may be given an expression such as $(x^2 - 144)$ and be asked to factorise it, therefore you need to recognise it as a difference of two squares.

In other words, x^2 is a square and 144 is also a square.

The answer here is $(x + 12)(x - 12)$.

Q Use this method to expand and simplify:

$(x + 7)(x - 7)$

PRACTICE

Expand and simplify:

1 $(x + 3)(x + 1)$ **2** $(x + 7)(x + 2)$ **3** $(x + 2)^2$

4 $(x + m)(x + n)$ **5** $(x + 4)(x + 2)$ **6** $(x - a)^2$

7 $(3 - x)^2$ **8** $(a - x)^2$ **9** $(x - y)^2$

10 $(x + 2)(x - 2)$ **11** $(x + 3y)(x - 3y)$ **12** $(6 - x)(x + 6)$

Regions

THE BARE BONES

➤ You have seen inequalities on a number line, you can also view them in regions.

➤ A region is an area of 2-D space with x and y axes, which is used to represent and illustrate the meaning of an inequality.

A Shaded and unshaded regions

KEY FACT

You shade out one region, so that the <u>area you require is the other non-shaded region</u>.

1 The graph shows the region $x > 4$.

2 This is the unshaded part on the right of the line $x = 4$.

3 Every point to the right of the line $x = 4$ satisfies this inequality.

4 The region that is shaded is the unwanted part for $x > 4$.

Q What inequality would describe the shaded region?

5 The line is dotted to show that $x = 4$ is not part of the required region.

B Which side of the line satisfies the inequality?

Q Can you use inequalities to describe the rectangle with vertices at (2, -1), (2, 5), (4, 5) and (4, -1)?

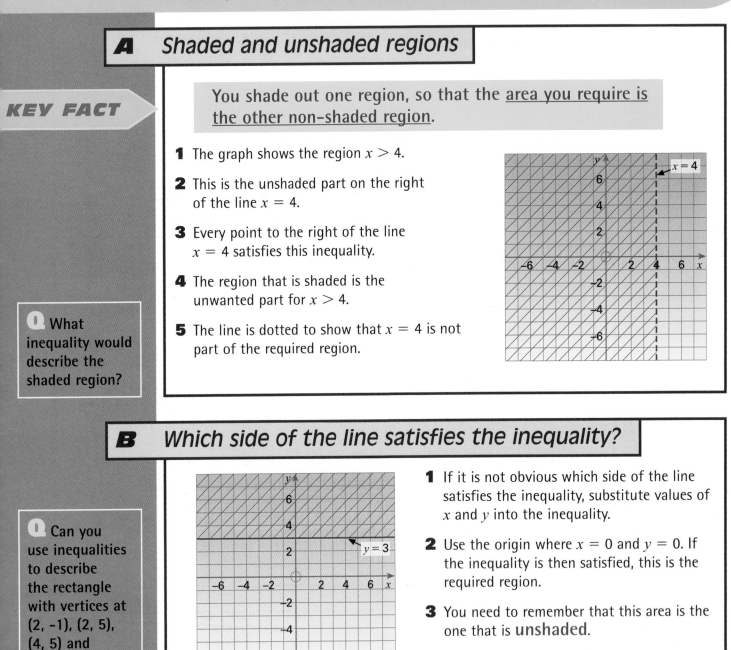

1 If it is not obvious which side of the line satisfies the inequality, substitute values of x and y into the inequality.

2 Use the origin where $x = 0$ and $y = 0$. If the inequality is then satisfied, this is the required region.

3 You need to remember that this area is the one that is **unshaded**.

4 This graph shows $y \leqslant 3$.

KEY FACT

Here the inequality is '<u>less than or equal to</u>', so the line is <u>included</u>. This is the reason why the <u>line is solid</u>.

c Some examples of inequalities and regions

1 The graph shows the region $y < 4x$.

- This line passes through the origin, so you need to test this region by using a different point.

- Substitute the values of x and y, for instance using (1, 1), to see if the equality $y < 4x$ is satisfied.

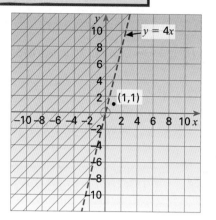

- Letting $x = 1$ and $y = 1$ does satisfy the inequality $y < 4x$. So the point (1, 1) does lie in the required region.

2

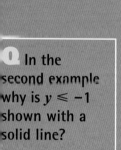

- Here $y \leqslant -1$ is shown with a solid line.
- So the region you want is on or below $y = -1$.

Remember
Lines with negative gradients slope from top left to bottom right.

3 Here $y < 5 - x$.

- Look at the line and the region that is shaded. Again this is the region you **do not** want.

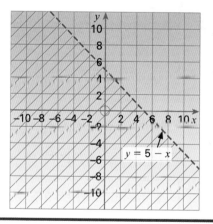

Q In the second example why is $y \leqslant -1$ shown with a solid line?

PRACTICE

Draw diagrams to show the regions that satisfy the inequalities:

1 $x < 3$ **2** $y \leqslant 5$

3 $x \geqslant 1$ **4** $-3 \leqslant y \leqslant 3$

5 $y \geqslant 3 - \frac{x}{2}$ **6** $y \geqslant 3x$

7 $y \leqslant 3x + 2$ **8** $3x - y > 1$

9 $x + 3y > 10$ **10** $x + y \geqslant 1$

Sequences and formulae

THE BARE BONES

➤ You can use a formula to find any term in a sequence.
➤ Given the sequence, you can work backwards and find the formula.
➤ The nth term is usually written u_n.
➤ An identity is an algebraic expression where the equality is true for any values of the variable.

A Finding the terms of a sequence

1 The nth term of a sequence is given by u_n, e.g. $u_n = 2n + 1$.

- Find the first two terms of this sequence:

  Substituting $n = 1$ the first term becomes $u_1 = (2 \times 1) + 1 = 3$

  Substituting $n = 2$ $u_2 = (2 \times 2) + 1 = 5$

- Therefore, the first two terms are 3 and 5.

- Note: $n = 1$ refers to the first term in the sequence and not the value for the first term. Here it shows that term one = 3.

KEY FACT

> n is called the position of a term in its sequence.

2 For the sequence: $u_n = 12 - 3n$, find u_2:

- Substituting $n = 2$, the second term becomes:
  $u_2 = 12 - (3 \times 2) = 6$

- So u_2 is 6.

3 For the sequence: $u_n = 3n^2 + 4$, find the third term in the sequence:

- Substituting $n = 3$, gives:
  $3(3^2) + 4 = 31$

- So the third term is 31.

Q What would the third term in the sequence $u_n = 2n + 1$ be?

B Finding which term has a given value

Remember
You will either need to find terms of a sequence or the formula for the sequence.

1 The number 64 is a term in a sequence given by the formula:

$$u_n = 5n + 4$$

2 Which term in the sequence has that value?

3 Substituting $u_n = 64$ into the formula, gives:

$$64 = 5n + 4$$
$$60 = 5n$$
$$n = 12$$

4 So 64 is the 12th term in the sequence.

Q Find the 100th term in the sequence.

C Finding the formula for a sequence

Q Find the formula for the sequence that begins 4, 9, 14, 19, 24 . . .

1 The first five terms of a sequence are:

7   11   15   19   23

2 Find the nth term.

3 Find the first differences using the table opposite.

position, n	①	②	③	④	⑤
terms, u_n	7	11	15	19	23
differences		+4	+4	+4	+4

The terms go up in 4s, so the nth term will be n lots of 4 ($4n$).

But the first term is 7, which is 3 more than $4n$.

When the first differences are all equal to a number, a, the formula is $u_n = an + b$, for a suitable value of b.

4 This suggests a formula of $u_n = 4n + 3$. Test it to see if it works.

D Identities

Remember
An equation is only true for one value of x.

1 The right-hand side of the equals sign in an identity is just another way of expressing what is written on the left-hand side of the equation.

2 $(x + 3)^2 = x^2 + 6x + 9$

- Substitute 1 into the left-hand side of the equation, you will find:

  LHS $= (1 + 3)^2 = 16$

- Now do the same to the right-hand side of the equation $1^2 + 6(1) + 9 = 16$

- So both sides are equal. In fact it doesn't matter what value you put in for x, you will find that LHS = RHS.

Q Check to see if the identities are equivalent: $3x^2 + 4$ and $x^2 + 3x + 2$.

- Strictly speaking, you should use $\equiv$ for 'equivalent to', so $(x + 3)^2 \equiv x^2 + 6x + 9$. However, you will often see the $=$ sign used instead of the $\equiv$ sign.

PRACTICE

1 Find the first four terms in these sequences:

(a) $u_n = 3n$ (b) $u_n = 2n + 1$

(c) $u_n = 4n - 7$ (d) $u_n = 17 - 2n$

(e) $u_n = 3^n$ (f) $u_n = \frac{2n}{n + 1}$

2 Find the value for n, for which u_n has the given value:

(a) $u_n = 5n + 12$ $u_n = 47$

(b) $u_n = 8(n - 3)$ $u_n = 32$

(c) $u_n = 2n^2 - 4$ $u_n = 158$

3 Find formula for u_n to describe each of these sequences:

(a) 2, 7, 12, 17, . . . (b) 6, 11, 16, 21, . . .

(c) 9, 17, 25, 33, . . . (d) 2, 5, 8, 11, . . .

More on formulae

➤ Manipulating formulae is an important skill used in solving equations and solving problems.

➤ Changing the subject means rearranging a formula to put one unknown on one side and everything else on the other side.

A The difference between a formula and an equation

1 Formulae:

$v = \pi r^2 h$ is a formula.

When you substitute in any value of r and h, you can calculate a corresponding value of v. In the same way, if you substitute any value of v, you can find a corresponding value for either r or h.

Look at this formula:

$v = u^2 + 2as$

Find the value of v when $u = 1$, $a = 2$, $s = \frac{1}{2}$

$v = 1^2 + 2(2)(\frac{1}{2})$

So $v = 3$

> The correct plural for formula is 'formulae', but it also may be written as 'formulas'.

KEY FACTS

> A formula applies to an infinite set of data, and can be used to calculate the corresponding value for any given value.

> Equations will only hold true for a particular value.

2 Equations:

$v = \frac{5}{3}\pi$ is an equation, it is only true for a **particular value** of v ($v = 5.24$ to 2 decimal places).

Q In the second example, if $y = +17$, what would the value of x be?

- In the equation, $y = 7x - 3$, find the value of y when $x = -2$

  $y = 7(-2) - 3$

  $y = -17$

- $t = \sqrt{\frac{m}{\pi}}$

  Find t when $m = 10$

  $t = \sqrt{\frac{10}{\pi}}$

  $t = 1.78$ (2 decimal places)

B Manipulating formulae

Sometimes you have to rearrange formulae.

1 Given $v = u + at$, find t when $v = 40$, $u = 5$ and $a = 6$.

Substitute in: $40 = 5 + 6t$

$6t = 35$

$t = 5.83$ (to 2 decimal places)

2 Given $u = 3(x + y)$, find u when $x = 5$ and $y = 7$.

$u = 3(5 + 7)$

$u = 36$

3 Given $m = \sqrt{\frac{p}{2\pi}}$, find p when $m = 100$

Square both sides to remove the square root sign:

$m^2 = \frac{p}{2\pi}$

Now rearrange the formula to make p the subject – multiply both sides by 2π and swap sides:

$p = 2\pi m^2$

Now substitute the given value of m:

$p = 2\pi \times 100^2$

$p = 62\,831.85$ (2 decimal places)

PRACTICE

1 $F = 2(a + b)$ Find F when $a = 4$ and $b = 7$.

2 $A = \pi r^2$ Find A when $r = 7.9$ cm (it is area).

3 $c = \pi d$ Find c when $d = 9.3$ cm.

4 $t = \sqrt{\frac{v}{4\pi}}$ Find v when $t = 12.49$.

5 $r = \sqrt{\frac{v}{\pi}}$ Find v when $r = 40$.

6 $v = 2\pi\sqrt{\frac{t}{g}}$ Find v when $t = 12$ and $g = 9.8$.

7 $s = \frac{1}{2}at^2$ Find t when $s = 100$ and $a = 10$.

8 $t = \frac{1}{2}m^2 n$ Find m when $t = 144$ and $n = 7$.

THE BARE BONES

➤ Linked quantities may conform to a number of relationships: for example, direct proportion, where one quantity is a multiple of another, or inverse proportion, where the product of the quantities is fixed.

➤ One quantity may be proportional or inversely proportional to a power of the other.

A Direct proportion

1 Suppose that two quantities y and x are in direct proportion.

> 'y is proportional to x' is written $y \propto x$.

If you plot them against each other on a graph, they form a straight line through the origin.

> If $y \propto x$, then $y = kx$ for some number k, called the constant of proportionality.

2 If two quantities are in direct proportion, and you know a pair of corresponding values, you can find the constant of proportionality and establish the equation linking them.

y is proportional to x. When $x = 250$, $y = 6.25$. What is the equation of proportionality?

$y \propto x$, so $y = kx$.

Substituting the known values gives $6.25 = k \times 250$, so $k = \frac{6.25}{250} = 0.025$ or $\frac{1}{40}$.

The equation of proportionality is therefore $y = 0.025x$ or $y = \frac{x}{40}$.

3 Note that other linear equations with a constant term (eg. $y = 2x + 5$) do not describe proportionality. This is because, for example, doubling x does not result in a doubling of y.

B Inverse proportion

1 Two quantities are in inverse proportion if, when one is multiplied by a number, the other is divided by that number (eg. if one is doubled, the other is halved).

> A relationship of this type is written $y \propto \frac{1}{x}$.

2 The equation of inverse proportionality is of the form $y = \frac{k}{x}$.

3 y is inversely proportional to x. When $x = 30$, $y = 0.6$.

Find the equation of proportionality, and the value of y when $x = 6$. $y \propto \frac{1}{x}$ so $y = \frac{k}{x}$.

Substituting the known values gives $30 = \frac{k}{0.6}$, so $k = 30 \times 0.6 = 18$.

The equation of proportionality is therefore $y = \frac{18}{x}$.

When $x = 6$, $y = \frac{18}{x} = \frac{18}{6} = 3$.

KEY FACTS

Remember
If two quantities are in direct proportion, then when one is multiplied/divided by a number, so is the other.

Q What is the value of x that will make y equal 60?

Q There is another way of writing the equation of proportionality, involving the product of the two quantities. Can you write it down?

C   Other types of proportionality

1 There are other types of proportionality involving other powers of x:

y is proportional to x	$y \propto x$	$y = kx$
y is inversely proportional to x	$y \propto \dfrac{1}{x}$	$y = \dfrac{k}{x}$
y is proportional to x^2	$y \propto x^2$	$y = kx^2$
y is inversely proportional to x^2	$y \propto \dfrac{1}{x^2}$	$y = \dfrac{k}{x^2}$
y is proportional to x^3	$y \propto x^3$	$y = kx^3$

2 The procedure for dealing with these relationships is exactly as for direct and inverse proportion: substitute known values, find k, then write down the equation of proportionality. This can then be used to answer any additional questions.

3 y is proportional to x^2. When $x = 4$, $y = 200$. What is the equation of proportionality, and what is the value of y when $x = 2$?

$y \propto x^2$, so $y = kx^2$.

Substituting the known values gives $200 = k \times 4^2$, so $16k = 200$ and $k = 12.5$.

The equation of proportionality is therefore $y = 12.5x^2$.

When $x = 2$, $y = 12.5x^2 = 12.5 \times 2^2 = 12.5 \times 4 = 50$.

Remember
Where the graph is a straight line, you can say that one quantity is directly proportional to the other.

Q Given the same starting values of x and y, repeat the example, given that $y \propto x^3$.

If a proportionality question asks you to 'establish a rule linking...' or 'find the relationship between...', you are being asked to find the equation of proportionality.

PRACTICE

1 The braking distance (y metres) of a car is proportional to the square of its speed (x km/h). Find the equation connecting x and y, and complete the table.

speed (x km/h)			64	80
braking distance (y metres)	10	24		

2 The distances of the planets from the Sun are often measured in astronomical units (au). 1au is defined to be the distance of the earth from the Sun (about 150 million km).

The speed (v km/sec) of a planet in its orbit is inversely proportional to the square root of its orbital distance (d au) from the Sun.

Find the rule linking d and v, and complete the table.

planet	Mercury	Earth	Neptune
orbital distance (d au)		1	30
speed (v km/s)	48		5

Quadratic functions

A Graphing quadratic functions

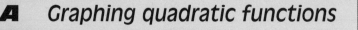

1 Quadratic expressions are terms that contain a squared term, such as $x^2 - 5$ or $3x^2 + 2x + 5$.
$\frac{1}{2}t(t - 5)$ is also a quadratic because on expansion, you get: $\frac{1}{2}t^2 - 2\frac{1}{2}t$.

2 Draw the graph of $y = x^2$. First make up a table of values.

KEY FACT

> To calculate the values for the table, substitute the values for x into the equation, to find the values of y.

x	-3	-2	-1	0	1	2	3
y	9	4	1	0	1	4	9

Then draw the graph using these points.

Notice that there is a line of symmetry in the graph.

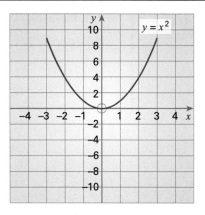

3 Draw the graph of $y = 2x^2$, taking values of x from -3 to $+3$.

Draw up the table:

x	-3	-2	-1	0	1	2	3
x^2	9	4	1	0	1	4	9
y	18	8	2	0	2	8	18

Then draw the graph using these points:

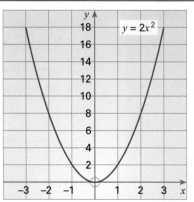

Q Where is the line of symmetry on this graph?

B Some further examples

1 Draw the graph of $y = 2x^2 + 5$ for values of x from -4 to $+4$.

Draw up a table of values:

x	-4	-3	-2	-1	0	1	2	3	4
x^2	16	9	4	1	0	1	4	9	16
$2x^2$	32	18	8	2	0	2	8	18	32
$+5$	$+5$	$+5$	$+5$	$+5$	$+5$	$+5$	$+5$	$+5$	$+5$
y	37	23	13	7	5	7	13	23	37

Remember
The x is the value that is squared, not the 2, so $4^2 = 16$, then double this to get 32 as shown in the table.

Draw the graph:

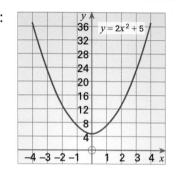

2 Draw the graph of $y = x^2 - 2x - 5$ for values of x from -4 to $+4$.

Draw up a table of values:

x	-4	-3	-2	-1	0	1	2	3	4
x^2	16	9	4	1	0	1	4	9	16
$-2x$	$+8$	$+6$	$+4$	$+2$	0	-2	-4	-6	-8
-5	-5	-5	-5	-5	-5	-5	-5	-5	-5
y	19	10	3	-2	-5	-6	-5	-2	3

Draw the graph:

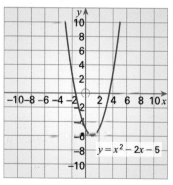

To draw a good graph:

- make sure you have enough points, usually about 8 to 10
- look at the range of values on the x-axis and use this as an indicator on where to site the y-axis, and vice versa
- sketch the curve to get an idea of what it looks like
- draw the curve, put your hand inside the curve to help you get a smooth curve, or draw round a flexi-curve.

Q If you have a graphical calculator, use it to graph these equations.

PRACTICE

1 Draw the graph of $y = x^2 + 4x + 1$, for suitable values of x.

2 Plot the curve of $y = 3x^2 - 8x + 4$.

3 Draw the graph of $y = \dfrac{x(x + 1)}{2}$. Where does the graph meet the y-axis?

THE BARE BONES

➤ A cubic function is one in which the highest power of x is 3.
➤ Curves of cubic functions usually have a turn in them.
➤ The reciprocal function is $y = \frac{1}{x}$.

A **Drawing up tables and graphs**

1 The simplest cubic function is $y = x^3$.

First draw up the table of values:

x	-4	-3	-2	-1	0	1	2	3	4
y	-64	-27	-8	-1	0	1	8	27	64

The cube of a negative number, will itself be negative.

Remember
When you draw
a curve, keep
your hand on
the inside of the
curve to get a
smoother finish,
or use a flexi–
curve.

Now draw the graph:

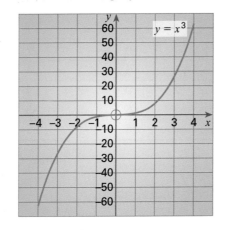

2 Now draw the graph:

Draw the graph of the curve described
by the equation $y = x^3 + 6x - 4$.
Draw up the table of values:

x	-3	-2	-1	0	1	2	3
y	-49	-24	-11	-4	3	16	41

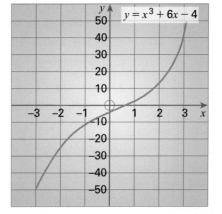

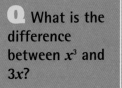

Q What is the
difference
between x^3 and
$3x$?

When the coefficient of x^3 is **positive**, the curve can be like the one below. Graphs of the form $y = ax^3$ all have a similar shape to this. When the value of a increases, the shape gets steeper.

When the coefficient of x^3 is **negative**, the curve can be like the one below. Graphs of the form $y = -ax^3$, where the value of a is negative, have a similar shape as when a is positive but reflected in the y-axis.

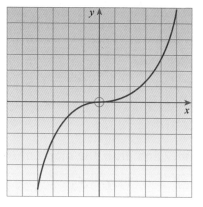

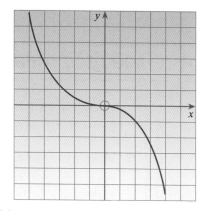

Graphs of the form $y = x^3 + c$ all have a similar shape to the one above. They all cut the y-axis at $(0, c)$.

More complicated cubic equations can produce graphs with two turning points, such as $y = x^3 - 3x - 2$.

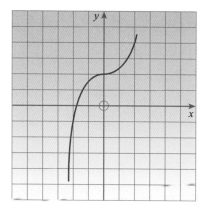

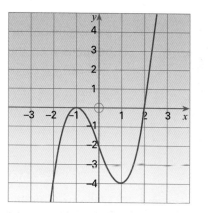

Q What does the graph of $y = x^3 + 4$ look like?

Memorise these curve shapes. You might need to recognise them in the exam.

PRACTICE

Draw the graphs of the following equations:

1 $y = 2x^3$

2 $y = -2x^3$

3 $y = x^3 + x^2$

4 $y = \frac{x^3}{2}$

5 $y = x^3 + 7x$

6 $y = 2x^3 + 7x$

7 $y = 2x^3 - 5x - 5$

8 $a = b^3 + 2b^3 + 3$

Graphing reciprocal functions

THE BARE BONES

➤ The reciprocal of a number is 1 divided by that number: reciprocal of x is $\frac{1}{x}$.

➤ Functions can be combined to form equations like:
$$y = ax^3 + bx^2 + cx + d + \frac{e}{x}$$

A   The reciprocal curve

KEY FACT

The reciprocal of a number is found by dividing 1 by that number, so the reciprocal of 2 is $\frac{1}{2}$.

1 Make up a table of values for $y = \frac{1}{x}$ and draw the graph of the equation.

Remember
Plot the axis carefully.

x	-5	-2	-1	$-\frac{1}{2}$	$-\frac{1}{5}$	$\frac{1}{5}$	$\frac{1}{2}$	1	2	5
y	$-\frac{1}{5}$	$-\frac{1}{2}$	-1	-2	-5	5	2	1	$\frac{1}{2}$	$\frac{1}{5}$

2 When the value of x is close to zero, the value of y gets very large. This means that the curve approaches the y-axis but never reaches it. The same thing happens with the x-axis when x is very large.

KEY FACT

There is no zero in the table because you cannot divide one by zero.

3 Make up a table of values for $y = 4 - \frac{2}{x}$ and draw the graph of the equation.

x	-3	-2	-1	$-\frac{1}{2}$	$-\frac{1}{5}$	$\frac{1}{5}$	$\frac{1}{2}$	1	2	3
y	4.7	5	6	8	14	-6	0	2	3	3.3

Q What happens when you multiply a number and its reciprocal together? Does this always happen?

B   Graphs of the form $y = ax^3 + bx^2 + cx + d + \frac{e}{x}$

1 There are times when you will have to graph equations that combine quadratics, cubics, linear and reciprocal functions.

2 When you have to draw them, you will usually find that a minimum of 2 of the coefficients will be zero.

3 Draw the graph of $y = 2x^2 - \frac{1}{x}$. As usual, draw a table of values.

x	-3	-2	-1	$-\frac{1}{2}$	$-\frac{1}{5}$	$\frac{1}{5}$	$\frac{1}{2}$	1	2	3
y	18.3	8.5	3	2.5	5.1	-4.9	-1.5	1	7.5	17.7

Now draw the graph, using the data from the table of values.

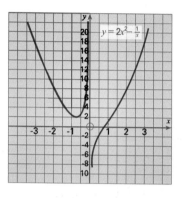

4 Draw the graph of $y = 2x^2 + \frac{1}{x}$. Again, you need a table of values.

x	-3	-2	-1	$-\frac{1}{2}$	$-\frac{1}{5}$	$\frac{1}{5}$	$\frac{1}{2}$	1	2	3
y	17.7	7.5	1	-1.5	-4.9	5.1	2.5	3	8.5	18.3

Now draw the graph.

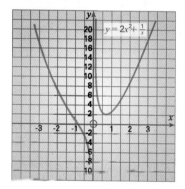

Q What is the effect of making a coefficient zero?

PRACTICE

Draw the graphs of the following equations:

1 $y = 4x^3 + 28x - 41$

2 $y = m + \frac{m^3}{10}$

3 $y = 3x^3 + 2x$

4 $t = 4s^2 + 3s + \frac{1}{s}$

5 $a = b^3 + \frac{b}{4}$

6 $y = m - \frac{m^3}{20}$

Solving equations with graphs

THE BARE BONES

➤ Accurate graphs can provide solutions to some equations and approximate solutions to others.

➤ Algebraic skills, such as expansion of brackets, are vital.

A Graphs and equations

1 You can use accurately drawn graphs to find approximate solutions to many equations.

KEY FACT

> The solutions of a quadratic equation are the values of x where the curve cuts the x-axis.

2 Draw the graph of $y = x^2 - 9$.

Draw up the table of values.

Remember
You know it is a quadratic, so it will be a parabola.

x	−4	−3	−2	−1	0	1	2	3	4
y	7	0	−5	−8	−9	−8	−5	0	7

Now draw the graph.

> Look at where the curve cuts the x-axis. It cuts at +3 and −3. These are the solutions to the equation.

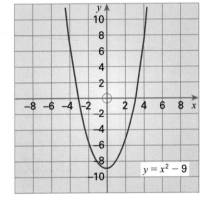

$y = x^2 - 9$

KEY FACT

Remember
The solutions to the equation are the values of x, where the curve cuts the x-axis.

3 Draw a graph of the equation $y = x^2 - 3x - 3$ for values of x from −5 to +5, use your graph to solve the equation $x^2 - 3x - 3 = 0$.

Draw up a table of values.

x	−5	−4	−3	−2	−1	0	1	2	3	4	5
y	37	25	15	7	1	−3	−5	−5	−3	1	7

Now draw the graph.

$y = x^2 - 3x - 3$

From the graph, you can see that the solutions of the equation are approximately −0.85 and 3.8. Try substituting these values into your equation $x^2 - 3x - 3 = 0$.

Q Where is the turning point of the second graph?

B Some examples

1 Draw the graph of $y = x^2 - 5x - 3$, use your graph to solve the equation $x^2 - 5x - 3 = 0$.

Draw up the table of values:

x	-3	-2	-1	0	1	2	3	4	5	6	7
x^2	9	4	1	0	1	4	9	16	25	36	49
^-5x	+15	+10	+5	0	-5	-10	-15	-20	-25	-30	-35
-3	-3	-3	-3	-3	-3	-3	-3	-3	-3	-3	-3
y	21	11	3	-3	-7	-9	-9	-7	-3	3	11

Now draw the graph.
Read off the approximate solutions to
the equation $x^2 - 5x - 3 = 0$.

You should find they are approximately
-0.6 and 5.6.

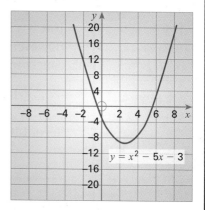

Remember
the graph is a
parabola.

2 Draw the graph of the equation
$y = x^2 - 4$ and hence, or otherwise, solve
the equation $x^2 - 4 = 6$.

First draw up the table of values:

x	-4	-3	-2	-1	0	1	2	3	4
y	12	5	0	-3	-4	-3	0	5	12

Draw the graph of: $y = x^2 - 4$, then add the
graph of $y = 6$.

Look at where the curve and the line cross,
read these values off the x axis.
These are the solutions to $x^2 - 4 = 0$.
You can see that the solutions for the equation are approximately $x = 3.2$ or -3.2.

Read off the
approximate
solutions to the
equation
$x^2 - 5x - 3 = 0$.
You should find
they are approx.
-0.6 and 5.6

PRACTICE

Draw graphs of the following:

1 $(x + 1)(x - 1)$ and use the graph to determine the solution to $x^2 - 1 = 3$.

2 $y = x^2 + 4$ and use the graph to find solutions of $x^2 + 4 = 7$.

3 $y = 2x^2 - 6x$ (use values of x from -1 to $+4$) and use your graph to find solutions
to the equation $2x^2 - 6x = 7$.

Trial and improvement

➤ Sometimes you can't find an exact solution to an equation, but can find a reasonable approximation using trial and improvement.

➤ There are two methods you can use: decimal search and bisection.

➤ The accuracy of the approximation depends on the number of trials: the more trials you do, the closer you can get to the solution.

A Decimal searches

1 Solve the equation $x^2 = 3$, correct to two decimal places.

This is a simple equation which you could solve easily using the square root key! However, it demonstrates the solution process clearly. Notice how the search concentrates on whole numbers, then decimals with one decimal place, then two.

KEY FACT

> Find the solution one significant figure at a time.

Record the results of the trials, together with your decisions, in a table like this:

x	x^2	Comments
1	1	Too small, so $x > 1$. Try $x = 2$.
2	4	Too big, so $x < 2$. x is between 1 and 2. Try numbers with one decimal place: try 1.5 first, as it's halfway between 1 and 2.
1.5	2.25	Too small, so $x > 1.5$. Try 1.7, (about halfway between 1.5 and 2).
1.7	2.89	Too small, so $x > 1.7$. Try 1.8.
1.8	3.24	Too big, so $x < 1.8$. x is between 1.7 and 1.8. Move on to numbers with two decimal places. Try 1.75, (between 1.7 and 1.8).
1.75	3.0625	Too big, so $x < 1.75$. Try 1.72 (between 1.7 and 1.75).
1.72	2.9584	Too small, so $x > 1.72$. Try 1.73.
1.73	2.9929	Too small, so $x > 1.73$. Try 1.74.
1.74	3.0276	Too big, so $x < 1.74$. x is between 1.73 and 1.74. You now need to know whether it's closer to 1.73 or 1.74. Trying 1.735 will decide.
1.735	3.010225	Too big, so $x < 1.735$. x is between 1.73 and 1.735.

Q Why was it necessary to try $x = 1.735$, when the answer only needs to be correct to 2 decimal places?

Therefore the solution is $x = 1.73$, to 2 dp.

KEY FACT

> You try out likely solutions in an equation to see how closely they fit. You use the results to make better guesses.
> That's why it's 'trial and improvement', not just 'trial and error'.

2 Solve the equation $p^3 - 10 = p$, correct to one decimal place.

It's much easier to 'hit the target' with trial and improvement if you're aiming for a fixed number, so rearrange the equation to read $p^3 - p = 10$.

A

p	$p^3 - p$	Comments
1	0	Too small, so $p > 1$. Try $p = 2$.
2	6	Too small, so $p > 2$. Try $p = 3$.
3	24	Too big, so $p < 3$. p is between 2 and 3. Try numbers with one decimal place: try 2.5 first, as it's halfway between 2 and 3.
2.5	13.125	Too big, so $p < 2.5$. Try 2.3, (about halfway between 2 and 2.5).
2.3	9.867	Too small, so $p > 2.3$. Try 2.4.
2.4	11.424	Too big, so $p < 2.4$. p is between 2.3 and 2.4. Is it closer to 2.3 or 2.4? Trying 2.35 will decide.
2.35	10.627875	Too big, so $p < 2.35$. Therefore the solution is $p = 2.3$, to 1 dp.

As you might have guessed, the solution ($p = 2.3089...$) is very close to 2.3!

B *Bisection*

1 In this method, you 'cut' **exactly** halfway between the previous two guesses.

Bisection generates accurate solutions quickly but is not 'user-friendly' because the trial numbers need to be carefully calculated and involve lots of decimal places early on.

2 Solve the equation $x^2 = 3$, correct to two decimal places, using bisection.

x	x^2	Comments
1	1	Too small, so $x > 1$. Try $x = 2$.
2	4	Too big, so $x < 2$. 1.5 is halfway between 1 and 2.
1.5	2.25	Too small, so $x > 1.5$. Try 1.75.
1.75	3.0625	Too big, so $x < 1.75$. Try 1.725.
1.725	2.975625	Too small, so $x > 1.725$. Try 1.7375.
1.7375	3.01890625	Too big, so $x < 1.7375$. Try 1.73125.
1.73125	2.9972265625	Too small, so $x > 1.73125$. Try 1.734375.
1.734375	3.008056640625	Too big, so $x < 1.734375$. x is between 1.73125 and 1.734375.

Therefore the solution is $x = 1.73$, to 2 dp.

Q Why was it necessary to go as far as $x = 1.734375$ in the bisection?

After each trial, write down as much information as you can about how you choose the next trial value.

PRACTICE

Use trial and improvement to find solutions to the following equations. Your answers should be to one decimal place.

1 $x^3 - 2x^2 = 1$ **2** $x^3 + 4x - 18 = 0$ **3** $5y^2 - 17y = \frac{7}{4}$

4 $6m^2 - 9 = 6m$ **5** $5t + \frac{3}{2} = -2t^3$

More complicated indices

THE BARE BONES

➤ In the expression x^n, x is called the base and n is called the index.
➤ The plural of index is 'indices'.
➤ Zero index means the term is 1 and negative indices indicate fractions.

A Multiplying and dividing indices

1 You have looked at how to multiply numbers in index form, e.g. $5^3 \times 5^4 = 5^7$.

In the same way: $x^3 \times x^4 = x^7$

KEY FACT

In general, in algebra, the rule for multiplying algebraic expressions with powers is:
$$a^m \times a^n = a^{m+n}$$

Remember
Variables must have the same base if the rule is to be applied. So $a^2 \times a^3 = a^5$, but $a^2 \times b^3 \neq ab^5$.

2 Simplify $3x^2 \times 4x^3$.

$3x^2 \times 4x^3 = 3 \times x^2 \times 4 \times x^3$
$= 3 \times 4 \times x^2 \times x^3$
$= 12 \times x^{2+3}$
$= 12x^5$

Watch out for this rule – it is used widely in other areas of mathematics.

3 Simplify $5x^3y^6 \times 4xy^3 \times 3y$.

$5x^3y^6 \times 4xy^3 \times 3y = 5 \times 4 \times 3 \times x^3 \times x \times y^6 \times y^3 \times y$
$= 60 \times x^{3+1} \times y^{6+3+1}$
$= 60x^4y^{10}$

Q Where did the indices of 1 come from in the calculation?

B Multiplying out $(a^m)^n$

Q Simplify: $(m^3)^4$.

1 $(y^3)^2$ means $(y^3) \times (y^3)$ or $y^3 \times y^3$

So $(y^3)^2 = y^{3+3} = y^{3\times2} = y^6$

KEY FACT

In general, the rule for multiplying algebraic expressions involving powers is: $(a^m)^n = a^{m\times n} = a^{mn}$

C Dividing out $a^m \div a^n$

1 Earlier you looked out how to divide numbers in index form.

For instance, $7^9 \div 7^3 = 7^{9-3} = 7^6$

In the same way, $x^8 \div x^5 = x^{8-5} = x^3$

Q Can you explain why $y^3 \div y = y^2$?

2 A different way of writing this, which is also correct, would be:

$$\frac{x \times x \times x \times x \times x \times x \times x \times x}{x \times x \times x \times x \times x}$$

3 If you cancel this division, you will find it gives you the same answer.

KEY FACT

Algebraically, this is written as $a^m \div a^n = a^{m-n}$

D Zero and negative indices

1 Look at this row: $2^2 = 4$; $2^3 = 8$; $2^4 = 16$; $2^5 = 32$; $2^6 = 64$; $2^7 = 128$.

Working from right to left, as the index goes down 1, what happens to the value?

You can see that as the index goes down by 1, the value halves, so this must mean that $2^1 = 2$ and that $2^0 = 1$.

KEY FACT

Any number raised to the power zero is 1.

Q Using your calculator, raise any number to the power zero. What do you notice about the answer?

2 It is also possible to find the value of an expression to a negative power.

For example: $8^6 \div 8^9 = 8^{-3} = \frac{1}{8^3}$

3 In the same way, $m^0 \div m^3 = m^{0-3}$

So $\frac{m^0}{m^3} = m^{-3}$

Anything to the power of zero $= 1$, so $\frac{1}{m^3} = m^{-3}$

KEY FACT

In general, in algebra you can say $a^{-m} = \frac{1}{a^m}$.

From the argument here, you can say that a^{-m} means the reciprocal of a^m.

PRACTICE

Work out the following:

1 $d^4 \times d^5$　　　　**2** $4r^6 \times 3r^7$　　　　**3** $g^3 \times g^2$

4 $15f^2g^3 \times 2fg$　　**5** $x^9 \div x^2$　　　　**6** $t^5 \div t$

7 $16m^3n \div 2mn$　　**8** $x^2 \div x^3$　　　　**9** $12j^5 \div 2j^7$

10 $4(y^3)^2$　　　　　**11** $(ab^4)^3$　　　　**12** $(5x^2)^2 \div x^3$

THE BARE BONES
➤ It is essential that you understand the techniques used to solve equations where the unknown occurs as part of an index.
➤ You can often solve these equations by comparing indices.

A Comparing indices

Remember
$a^{-m} = \frac{1}{a^m}$

Remember
$(a^m)^n = a^{mn}$

Remember
You are using the rules of indices already discussed on the previous pages.

> KEY FACT

1 In the following example, you are trying to find the value of the index, which is called k.

$$x^k = \frac{1}{x^{-3}}$$

- Simplify the above by taking the denominator to the power –1:

$$x^k = (x^{-3})^{-1}$$

$$x^k = x^3$$

- By comparing indices, you find that $k = 3$.

> Comparing values in this way is a straightforward and useful technique.

2 Here you are trying to find the value of y.

$$x^y = \sqrt[3]{x} \div \frac{1}{x^4}$$

- Again, you need to simplify the equation.

$$= x^{\frac{1}{3}} \div x^{-4}$$

$$= x^{\frac{1}{3}} \times x^4$$

- So $x^y = x^{4\frac{1}{3}}$

- By comparing indices, you find that $y = 4\frac{1}{3}$.

Q Can you explain why $\div x^{-4}$ became $\times x^4$?

> KEY FACT

> Always write indices as fractions rather than decimal numbers. A fraction like $\frac{1}{3}$ is an exact number, the decimal 0.333 is an approximation.

A

3 Have a look at the following equation.

$$x^y = \frac{x^2 \times \sqrt{x^4}}{(\sqrt[4]{x})^3}$$

- To solve it, you need to find the value for the index y, on the left-hand side.
- Begin by simplifying the equation like this:

$$x^y = \frac{x^2 \times x^2}{x^{\frac{3}{4}}}$$

$$x^y = \frac{x^4}{x^{\frac{3}{4}}}$$

$$x^y = x^{(4 - \frac{3}{4})}$$

$$x^y = x^{3\frac{1}{4}}$$

- So by comparing indices, you can work out that $y = 3\frac{1}{4}$.

KEY FACT

Any value to the power zero equals 1.

PRACTICE

Remember
To be successful at higher tier GCSE, you have to show that you are good at using algebra. It is, therefore, essential to understand and be able to apply the different rules of indices.

Find the value of the letter given in brackets after each equation:

1 $x^y = \sqrt{x}$ $\qquad$ (y)

2 $z^k = 1$ $\qquad$ (k)

3 $4^y = 64$ $\qquad$ (y)

4 $x^{\frac{1}{2}} = 25$ $\qquad$ (x)

5 $3^k = 27$ $\qquad$ (k)

6 $m^k = m^{\frac{1}{2}} \div \sqrt{\frac{1}{m}}$ $\qquad$ (k)

7 $m^{(y + 1)} = (m^{-2})^3$ $\qquad$ (y)

You can use your calculator to check your answer. Choose a value for the letter you haven't found, substitute it and see if both sides of the equation match.

THE BARE BONES

> Quadratic expressions are expressions in which the highest power is 2.
> In general, they have the form $ax^2 + bx + c$.
> In the general form, the coefficient of x^2 is a, the coefficient of x is b and the constant term is c.

A Factorising quadratics of the form $ax^2 + bx$

Remember
HCF stands for highest common factor.

1 Factorise $4x^2 + 12x$.

2 First of all, look for the highest common factor of $4x^2$ and $12x$.

3 The HCF is the largest term that will go into $4x^2$ and $12x$, which is $4x$.

4 When $4x^2 + 12x$ is factorised, it becomes $4x(x + 3)$.

B Factorising quadratics of the form $x^2 + bx + c$

Remember
By factoring out the HCF, you are making the expression more manageable and therefore easier to factorise.

1 Factorise $x^2 - 9x + 14$.

2 First of all, check the expression is in the form $ax^2 + bx + c$.

3 Now multiply the coefficient of x^2 by the constant term: $a \times c = 14$.

4 Find two factors of this product which when added give the coefficient of x (-9). Here is a list:

14	7	-7	-14
1	2	-2	-1
15	9	-9	-15
		✓	

The factors are -2 and -7.

5 Now you need to rewrite the bx term using the two factors, in other words, $-2x$ and $-7x$.

6 Now continue as if you were factorising by pairing.

$$x^2 - 9x + 14 = x^2 - 2x - 7x + 14$$
$$= x(x - 2) - 7(x - 2)$$
$$= (x - 7)(x - 2)$$

Q Factorise $x^2 - 2x - 24$ by this method.

Write down a complete list of the factors in pairs. Don't forget that positive numbers can have negative factors.

c   Factorising quadratics: $ax^2 + bx + c$ where $a > 1$

Wherever possible, it is better to write the expression in the form $ax^2 + bx + c$.

Remember
The coefficient of x^2 is the number multiplying it.

1 Factorise $6y^2 + 42y + 60$.

2 Factorise out the HCF. The HCF is 6 so:

$6y^2 + 42y + 60 = 6(y^2 + 7y + 10)$

3 Multiply out the coefficient of y^2 (1) by the constant (10) and find two factors that when added come to 7, i.e 5 and 2.

4 Rewrite the $7y$ term using these two factors:

$= 6(y^2 + 5y + 2y + 10)$

Q Can you explain why there is a 6 outside the square brackets in this example?

5 You now need to factorise the bracket as you would for pairing:

$= 6[(y(y + 5) + 2(y + 5)]$

$= 6[(y + 2)(y + 5)]$

D   Factorising quadratics where x is negative

1 Factorise $18x - 12 - 6x^2$.
First remove the HCF from all of the terms.
$18x - 12 - 6x^2 = 6(3x - 2 - x^2)$
Rearrange this into the proper order:
$6(-x^2 + 3x - 2)$

2 To factorise the expression in the bracket, form the product $ac = -1 \times -2 = 2$.
So the factors of 2 that add to 3 are 1 and 2. Rewrite the middle term and proceed as for factorising by pairing.

$6(-x^2 + 3x - 2) = 6(-x^2 + 2x + x - 2)$
$= 6[-x(x - 2) + 1(x - 2)]$
$= 6(-x + 1)(x - 2)$

Q Can you explain why we have used square brackets here?

3 This is correct, but the first bracket looks neater if it doesn't start with a minus sign, so rearrange its terms:

$18x - 12 - 6x^2 = 6(1 - x)(x - 2)$.

Note that it is possible, by pairing the terms differently, to obtain $6(x - 1)(2 - x)$. This is also correct!

PRACTICE

Factorise:

1 (a) $x^2 + 4x$ (b) $9x + 18$ (c) $x^2 + 15x + 36$

2 (a) $x^2 + 6x$ (b) $12x - 144$ (c) $x^2 + 3x - 10$

3 (a) $x^2 - 8x + 16$ (b) $2x^2 - 6x + 4$ (c) $6x^2 - 42x + 72$

4 (a) $10x^2 + 20x + 10$ (b) $8a^2 + 16a + 8$ (c) $m^2 - 6m + 9$

The difference of two squares

THE BARE BONES

➤ A difference of two squares is an expression like $10^2 - 5^2$, $2.1^2 - 1.8^2$, $3^2 - 4^2$, etc.

➤ The general expression is usually written $a^2 - b^2$.

A *The difference of two squares*

1 The expression $10^2 - 5^2$ has the value $100 - 25 = 75$.

The expression $(10 - 5)(10 + 5)$ has the value $5 \times 15 = 75$.

It's no coincidence: this happens with any pair of numbers you try.

2 More examples:

$7^2 - 2^2 = 49 - 4 = 45 = 5 \times 9 = (7 - 2)(7 + 2)$

$2.1^2 - 1.8^2 = 4.41 - 3.24 = 1.17 = 0.3 \times 3.9 = (2.1 - 1.8)(2.1 + 1.8)$

$3^2 - 4^2 = 9 - 16 = -7 = -1 \times 7 = (3 - 4)(3 + 4)$

3 In general, using algebra, $a^2 - b^2 = (a - b)(a + b)$. This mathematical sentence is an **identity**, meaning that it is **always** true, whatever the values of a and b.

Q Try out some more numerical examples on your calculator to convince yourself that the identity is true.

B *Factorising using the difference of two squares*

1 To factorise an expression you think is a difference of two squares:

First check that the expression is a difference of two squares. Are both terms squares and is one subtracted from the other?

KEY FACT

'Difference' always involves subtraction.

2 Write the expression in the form $a^2 - b^2$. You now know what a and b represent in your expression.

3 Factorise into the form $(a - b)(a + b)$.

Example: factorise $x^2 - 4$.

Is it a difference of two squares? Yes, x^2 is the square of x and 4 is the square of 2. One is being subtracted from the other.

4 Write it in the correct form: $x^2 - 4 = x^2 - 2^2$. So a represents our x and b represents 2.

$(a - b)(a + b)$ will be written $(x - 2)(x + 2)$.

So $x^2 - 4 = (x - 2)(x + 2)$.

Q Expand the brackets on the right-hand side of the identity. Do you get the correct expression after you've simplified the expanded version?

B

Q Can you explain why $9y$ appears in this factorisation?

5 Example: factorise $16x^2 - 81y^2$

$16x^2$ is the square of $4x$.

$16x^2 - 81y^2 = (4x - 9y)(4x + 9y)$.

> The general expression for a difference of two squares is $a^2 - b^2$

C Harder examples with a common factor

Q Expand and simplify the final expression on the right. Satisfy yourself that it is the same as the original quadratic expression.

1 Consider an expression such as $48p^2 - 75q^2$. It looks a bit like a difference of two squares, but neither of the numbers is a square.

2 However, there is a common factor of 3 between the terms.

$48p^2 - 75q^2 = 3(16p^2 - 25q^2)$.

3 The expression inside the brackets *is* a difference of two squares.

So $16p^2 - 25q^2 = (4p - 5q)(4p + 5q)$, and, therefore:

$48p^2 - 75q^2 = 3(4p - 5q)(4p + 5q)$.

> Make sure you learn the squares of all of the numbers up to 25.

PRACTICE

Remember
When factorising, ask yourself: is it straightforward factorisation? Does the HCF need factorising out first? Is it a difference of two squares?

Decide if these expressions are a difference of two squares and if they are, factorise them.

1 $x^2 - y^2$

2 $4m^2 + 9y^2$

3 $49t^2 - 64m^2$

4 $9z - 16y$

5 $12m^2 - 16n^2$

6 $25h^2 + 36j^2$

7 $100k^2 - 81m^2$

8 $169c^2 - 225d^2$

9 $625k^2 - 400m^2$

10 $1600v^2 - 400u^2$

11 $18p^2 - 2q^2$

12 $k^2 - \dfrac{m^2}{4}$

13 $a^4 - b^4$

14 $G^2 - \dfrac{1}{H^2}$

15 $\dfrac{a^2}{3} - \dfrac{1}{48B^6}$

Algebraic fractions

> $\frac{3x^2}{6x}$ and $\frac{8(x+9)}{2x}$ are examples of algebraic fractions.
> Fractions can only be simplified by cancelling, when and only when there is a common factor in the numerator and the denominator.

A Simplifying fractions

1 Simplify $\frac{3x^4}{6x}$

$$\frac{3x^4}{6x} = \frac{3 \times x \times x \times x \times x}{6 \times x} = \frac{1 \times x \times x \times x}{2} = \frac{x^3}{2}$$

2 Simplify $\frac{8(x+9)}{2x}$

Look for the common factors. Obviously 8 and 2 will cancel because they have common factors, so: $\frac{8(x+9)}{2x} = \frac{4(x+9)}{x}$

> You cannot divide the x's in example 2, because x is not a common factor.

Q Can you explain what is meant by a common factor?

KEY FACT

B Multiplying algebraic fractions

KEY FACT

To multiply algebraic fractions:
Look for and cancel any factor that is common to the numerator and the denominator.
Only when it is not possible to cancel any further, multiply the remaining numerators and denominators.

Remember
Use the rules of indices to ensure that you have cancelled correctly.

1 Simplify $\frac{x^6}{y^2} \times \frac{x^2y}{z} \times \frac{z^2}{x^3}$

Cancel the x's first: x^6 and x^3 will cancel.

Then cancel the y^2 with the y, leaving y in the denominator.

Now cancel z^2 with z

So $\frac{x^6}{y^2} \times \frac{x^2y}{z} \times \frac{z^2}{x^3} = \frac{x^3}{y} \times x^2 \times z$

$= \frac{x^5z}{y}$

Q In this example, can you explain why there is not a z in the denominator of the answer?

C   Dividing algebraic fractions

When you divide by an algebraic fraction, it is the same as multiplying by its reciprocal.

1 Look at what happens when you work in numbers.

$4 \div \frac{1}{2}$ means how many $\frac{1}{2}$'s are there in 4?

Clearly there are eight halves in 4, so $4 \div \frac{1}{2} = 8$.

But the reciprocal of $\frac{1}{2}$ is 2.

Be very careful not to miss out any steps, as these are complicated manipulations. Check with your calculator, by substituting a value for x.

The reciprocal of a number is the number you multiply by to get an answer of 1.

2 $4y \div \frac{x}{z} = \frac{4y}{1} \times \frac{z}{x}$

$= \frac{4yz}{x}$

D   Adding and subtracting algebraic fractions

1 As with ordinary fractions, you can't add or subtract unless the fractions have a **common denominator**.

Simplify $\frac{3x + 1}{2} + \frac{2x - 3}{3}$

The denominators are 2 and 3, so the lowest common denominator (LCD) is 6.

The lowest common denominator is the LCM of the denominators of the original fractions.

2 Multiplying the top and bottom of the first fraction by 3, $\frac{3x + 1}{2} = \frac{3(3x + 1)}{6}$.

Multiplying the top and bottom of the second fraction by 2, $\frac{2x - 3}{3} = \frac{2(2x - 3)}{6}$.

So $\frac{3x + 1}{2} + \frac{2x - 3}{3} = \frac{3(3x + 1)}{6} + \frac{2(2x - 3)}{6}$.

$= \frac{9x + 3}{6} + \frac{4x - 6}{6}$

$= \frac{9x + 3 + 4x - 6}{6} = \frac{13x - 3}{6}$

Simplify the following:

1 $\frac{x^3y^2}{z^2} \times \frac{xyz}{x^4}$

2 $\frac{12mn}{10mn}$

3 $\frac{3(x + 4)}{6(x + 4)(x - 4)}$

4 $5x \div \frac{1}{x}$

5 $8mx^2 \div \frac{2}{mx}$

6 $\frac{2a + 3}{4} - \frac{a + 2}{5}$

7 $\frac{t^2 + 4t + 3}{3t} + \frac{t^2 - 2}{2t}$

Quadratic equations 1

➤ Equations of the form $ax^2 + bx + c = 0$ where $a \neq 0$ are called quadratic equations.
➤ Quadratic equations either have no roots, or two roots.
➤ There are three ways to solve equations of this type.

A Solving quadratic equations by factorising

KEY FACT

A quadratic equation of the form $ax^2 + bx = 0$, where $a \neq 0$, has two solutions, which are called roots. The roots may be equal.

1 Solving quadratics of the form $x^2 = bx$.

Solve the equation $x^2 = 12x$.

- The first step is to rearrange it into the form $ax^2 + bx + c = 0$.

- So the equation becomes $x^2 - 12x = 0$ ($c = 0$ in this case).

KEY FACT

If the product of two terms is zero, one of the terms must be zero

2 Now factorise and you get $x(x - 12) = 0$.

- Either $x = 0$ or $x - 12 = 0$.

- If $x - 12 = 0$ then $x = 12$.

3 So this means that $x = 0$ and $x = 12$ are the two solutions or roots of the equation $x^2 = 12x$.

You must avoid the common error of dividing both sides by x. This tactic loses the $x = 0$ solution. In other words, had you divided both sides by x, you would simply have $x = 12$ as the solution, and quadratics must have two solutions.

Q Solve $x^2 + 5x = 0$ by this method.

B Solving equations of the form $(ax + b)(cx + d) = e$

Remember
Rearrange the equation into the form $ax^2 + bx + c = 0$

1 Solve the equation $(2x + 1)(3x - 1) = 4$.

Expand the brackets $2x(3x - 1) + 1(3x - 1) = 4$

$6x^2 - 2x + 3x - 1 = 4$, so $6x^2 + x - 1 = 4$

Rearranging into the correct form, $6x^2 + x - 5 = 0$.

2 Now calculate ac (-30) and find factors of ac that add to b (1). They are 6 and -5.

$6x^2 + 6x - 5x - 5 = 0$.

$6x(x + 1) - 5(x + 1) = 0$.

3 So $(6x - 5)(x + 1) = 0$.

4 So either $6x - 5 = 0$ or $x + 1 = 0$.

5 So the two solutions are $x = \frac{5}{6}$ or $x = -1$.

Q Can you explain how we worked out line 5, to another person?

C Solving quadratic equations with repeated roots

Sometimes the two roots of a quadratic equation may be equal. In these cases it looks as if the equation only has one root, but in fact it has repeated roots.

1 Solve the equation $m^2 - 7m + 27 = 2 + 3m$.

2 This rearranges to give $m^2 - 10m + 25 = 0$.

3 Now you need factors of 25 (because $ac = 25$) that sum to -10 (because $b = -10$).

4 So rewrite for factorisation by pairing and you get:

$m^2 - 5m - 5m + 25 = 0$

$m(m - 5) - 5(m - 5) = 0$

$(m - 5)(m - 5) = 0$

5 Therefore, $m = 5$ (repeated).

> This is possible; it is called a repeated root.

Remember
Rearrange it into the form $am^2 + bm + c = 0$.

Q Can you draw the graph of this equation, to see the repeating root?

KEY FACT

Always check your solution by substituting into the original equation. Make sure you do this with both solutions.

D More solutions

1 Solve $2m^2 - 288 = 0$.

2 Take out the common factor.

$2(m^2 - 144) = 0$.

> You have seen this before, it is the difference of two squares.

3 Using the difference of two squares gives you:

$2(m - 12)(m + 12) = 0$.

4 For this equation to be true, either $m - 12 = 0$ or $m + 12 = 0$.

5 So $m = 12$ or -12.

KEY FACT

Q Can you explain why 144 is a square number and say what the next square number after 144 is?

PRACTICE

Solve the following equations.

1 $x^2 = 16x$

2 $t^2 = 20t$

3 $b^2 = 49b$

4 $x^2 + 8x + 12 = -4$

5 $(x + 2)(x + 3) = 0$

6 $m^2 + 14m + 56 = 7$

7 $y^2 - 20 = 5$

8 $4t^2 - 400 = 0$

9 $g^2 - 6g - 30 = 3g + 6$

10 $h^2 - 8h - 60 = 4h + 4$

Quadratic equations 2

➤ Equations of the type $y^2 = k$ are a special case.

➤ Completing the square is a technique you use to solve quadratics that cannot be factorised.

➤ There is a formula that can be used to solve quadratic equations.

A Equations of the type $y^2 = k$

To solve equations of this type, we need to take the square root of both sides, but it is vital to write the $\pm$ in front of the square root. This is to indicate that the negative root is also a valid answer.

KEY FACT

> By definition, the square root symbol indicates the positive square root only.

Q Can you explain why it is essential to have the negative square root?

1 Solve the equation $(4x - 2)^2 = 81$.

2 Take the square roots of both sides. This gives $4x - 2 = \pm 9$.

3 Now add 2 to both sides: $4x = 2 \pm 9$.

4 Either $4x = 11$ or $4x = -7$. Therefore, $x = \frac{11}{4}$ or $-\frac{7}{4}$.

B Solving a quadratic by completing the square

KEY FACT

> Not all quadratic expressions can be factorised into whole numbers, which means you cannot solve all quadratic equations by factorisation.

1 Solve $x^2 + 8x - 1 = 0$, leaving your answer in surd form.

2 Rewrite this in the form $ax^2 + bx = c$.
$x^2 + 8x = 1$.

3 Now add to both sides the square of half of the coefficient of x. This is to make the left-hand side of the equation into a squared term.

KEY FACT

> As long as both sides of the equations are treated in the same manner, the equation remains balanced and equal.

4 Since $b = 8$, $\left(\frac{b}{2}\right)^2 = \left(\frac{8}{2}\right)^2 = 4^2 = 16$

5 So you get $x^2 + 8x + 16 = 1 + 16$.

6 The left-hand side of the equation is now the expansion of $(x + 4)^2$.

7 So you can now rewrite the equation as $(x + 4)^2 = 17$.

You can solve it in the same way you solved the equation in section A; by taking the square roots of both sides of the equation. $x + 4 = \pm \sqrt{17}$.

Q Can you explain how you got line 6?

8 Either $x = 4 + \sqrt{17}$ or $x = 4 - \sqrt{17}$.
Since this is in surd form, you have the final solution.

c Solving a quadratic equation by formula

The formula for the solution of the quadratic equation $ax^2 + bx + c = 0$ is given as: $x = \dfrac{-b \pm \sqrt{b^2 - 4ac}}{2a}$

This formula gives the solutions for a quadratic equation with any given values for a, b, and c.

The equation MUST be arranged in standard order, before using the formula.

If $b^2 - 4ac$ is negative, the equation has no solution.

1 Solve $x^2 + 7x + 3 = 0$, giving your answers to two decimal places.

$$x = \frac{-b \pm \sqrt{b^2 - 4ac}}{2a}$$

$a = 1$, $b = 7$, $c = 3$, so $x = \dfrac{-7 \pm \sqrt{7^2 - (4 \times 3)}}{2}$

$$= \frac{-7 \pm \sqrt{37}}{2}$$

2 So $x = -0.46$ or $x = -6.54$, to 2 d.p.

The sum of the calculated roots should be equal to $-\dfrac{b}{a}$. This is a useful check on every question and you are encouraged to use it.

If you are asked to solve the equation by factorising and you do so by using the formula, you will receive no marks.

Solve to 2 decimal places:

1 $(x + 3)^2 = 10$

2 $(x + 1.5)(x + 1.5) = 0$

3 Solve $x^2 + 4x + 1 = 0$ by completing the square.

4 Solve $x^2 + 15x - 3.5 = 0$ by using the formula.

5 Try to solve the equation $x^2 - 3x + 4 = 0$. Explain why this is not possible and adjust answer accordingly.

THE BARE BONES

➤ Simultaneous equations in the exam will be one of two types: both linear or one linear and one quadratic.

➤ When one of the equations is quadratic, there will usually be two sets of solutions.

A Solving equations 1

1 Solve the equations:

$$x^2 + 2y^2 = 30$$
$$x + y = 5.$$

2 First of all, rewrite them and label them in the same way you do with simultaneous linear equations. So you have:

$$x^2 + 2y^2 = 30 \text{ (i)}$$
$$x + y = 5 \text{ (ii)}$$

Remember
Keep the equals sign aligned.

3 Now, because you have **two unknowns and one set of quadratics**, you need to rearrange equation (ii), either in terms of x or y and then substitute that value into equation (i).

4 Here you are going to rearrange equation (ii) to make y the subject, because this is the easier substitution. From equation (ii):

$$y = 5 - x \text{ (iii)}$$

Remember
Square the bracket first.

5 So now substitute for y into equation (i). This now becomes:

$$x^2 + 2(5 - x)^2 = 30$$
$$x^2 + 2[(5 - x)(5 - x)] = 30$$
$$x^2 + 2[25 - 10x + x^2] = 30$$
$$x^2 + 50 - 20x + 2x^2 = 30$$
$$3x^2 - 20x + 50 = 30$$
$$3x^2 - 20x + 20 = 0$$

6 Now you can solve this using the quadratic formula.

KEY FACT

> You need to memorise the formula and be able to quote it.

$$x = \frac{-b \pm \sqrt{b^2 - 4ac}}{2a}, \text{ where } a = 3, b = -20, c = 20$$

7 $x = \dfrac{20 \pm \sqrt{160}}{6}$, so $x = 1.23$ or $x = 5.44$, to 2 d.p.

8 When $x = 1.23$, $y = 3.77$, when $x = 5.44$, $y = -0.44$.

Q How were the values found in step 8?

B Solving equations 2

1 Solve the equations:

$$y = 5x^2 \text{ (i)}$$
$$x - y = 0 \text{ (ii)}$$

2 Rearrange equation (ii) to make x the subject.
$x = y$, so substituting this in equation (i), $y = 5y^2$.

3 Rearrange: $5y^2 - y = 0$.

4 Factorise: $y(5y - 1) = 0$.

5 So either $y = 0$ or $5y - 1 = 0$, $y = \frac{1}{5}$.

6 When $y = 0$, $x = 0$. When $y = \frac{1}{5}$, $x = \frac{1}{5}$.

Q Can you explain line 6 to another person?

C Repeated roots

1 Quadratic equations may have repeated roots, so some pairs of simultaneous equations have them, too.

2 Consider the equations

$$x^2 + y^2 = 25 \text{ (i)}$$
$$3x = 25 - 4y \text{ (ii)}$$

Make x the subject of (ii):
$$x = \frac{25 - 4y}{3}$$

3 Substitute this into (i): $\left(\frac{25 - 4y}{3}\right)^2 + y^2 = 25$

So $\frac{(25 - 4y)^2}{9} + y^2 = 25$

Multiply by 9: $(25 - 4y)^2 + 9y^2 = 225$

Expand: $625 - 200y + 16y^2 + 9y^2 = 225$

So: $25y^2 - 200y + 625 = 225$

and $25y^2 - 200y + 400 = 0$

Divide by 25: $y^2 - 8y + 16 = 0$

Factorise: $(y - 4)^2 = 0$

So there is a repeated root, $y = 4$.

4 If $y = 4$, then $3x = 25 - 16 = 9$, so $x = 3$.

The solution is $x = 3$, $y = 4$, repeated.

> Always remember to check both sets of solutions by substitution. That's all it takes to discover an error that might cost several marks.

Q What does the repeated root mean in terms of the graphs of the equations?

PRACTICE

Solve the following, if possible:

1 $x^2 + y^2 = 25$
  $x + y = 7$

2 $xy = 5$
  $2x - y = -3$

3 $y^2 = 2x$
  $x - y = 0$

4 $x^2 - y^2 = 20$
  $x + y = 7$

5 $2x^2 + 5y = 38$
  $x + y = 7$

Congruence and similarity

➤ Congruent 2-D shapes are shapes that are identical.

➤ Shapes that are mathematically similar are identical, except in size.

➤ In exam questions, congruent and similar shapes are not usually drawn accurately.

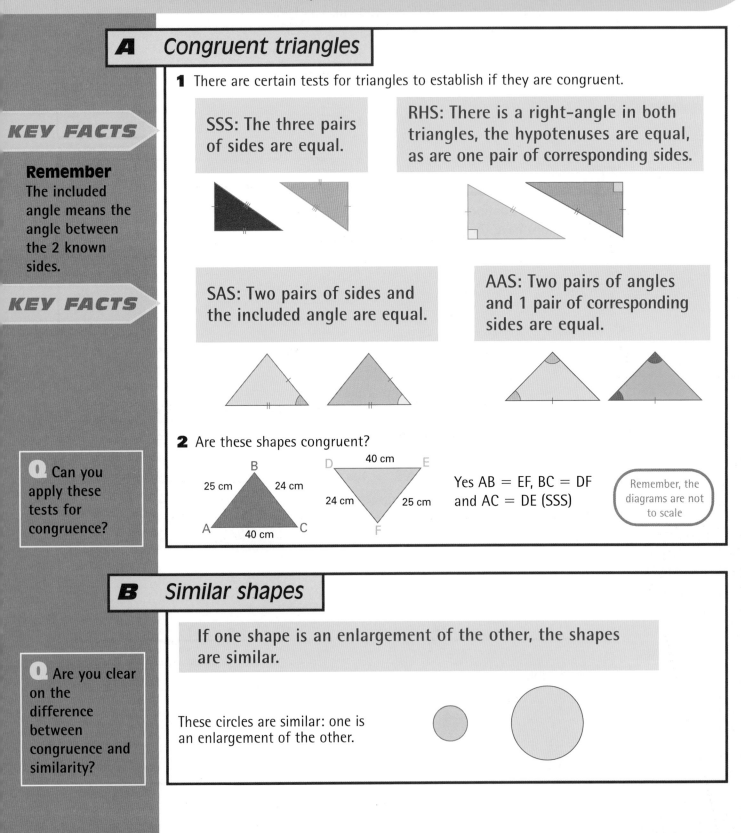

A Congruent triangles

KEY FACTS

Remember
The included angle means the angle between the 2 known sides.

KEY FACTS

1 There are certain tests for triangles to establish if they are congruent.

SSS: The three pairs of sides are equal.

RHS: There is a right-angle in both triangles, the hypotenuses are equal, as are one pair of corresponding sides.

SAS: Two pairs of sides and the included angle are equal.

AAS: Two pairs of angles and 1 pair of corresponding sides are equal.

2 Are these shapes congruent?

B
25 cm   24 cm
A   40 cm   C

D   40 cm   E
24 cm   25 cm
F

Yes AB = EF, BC = DF and AC = DE (SSS)

Remember, the diagrams are not to scale

Q Can you apply these tests for congruence?

B Similar shapes

If one shape is an enlargement of the other, the shapes are similar.

These circles are similar: one is an enlargement of the other.

Q Are you clear on the difference between congruence and similarity?

C   The scale factor of enlargement

> **The scale factor of enlargement is the ratio:**
>
> $$\frac{\text{side length of image shape}}{\text{corresponding side length of object shape}}$$

Y FACT

1 These rectangles are similar. Find the length (x) of the second rectangle.

2 Here you use the ratio of the corresponding sides, i.e. $\frac{11}{5} = 2.2$

3 So the scale factor of enlargement is 2.2.

4 Now to find x, simply work out 2.2×9.

5 $2.2 \times 9 = 19.8$ cm, so $x = 19.8$ cm.

Q Can you explain why we calculate $\frac{11}{5}$ rather than $\frac{5}{11}$?

D   Similarity in triangles

Two triangles are similar if one of the following is true:

    (a) all corresponding angles are equal.

    (b) all corresponding sides are in the same ratio

    (c) 2 pairs of corresponding sides are in the same ratio and the included angle is equal.

1 Find the length marked x.

2 The triangles have 2 pairs of equal angles, therefore the third pair must be equal.

3 The scale factor of enlargement is $\frac{9}{6} = 1.5$.

4 So $x = 1.5 \times 11.8 = 17.7$ cm.

Q Can you explain line 3 in the calculation?

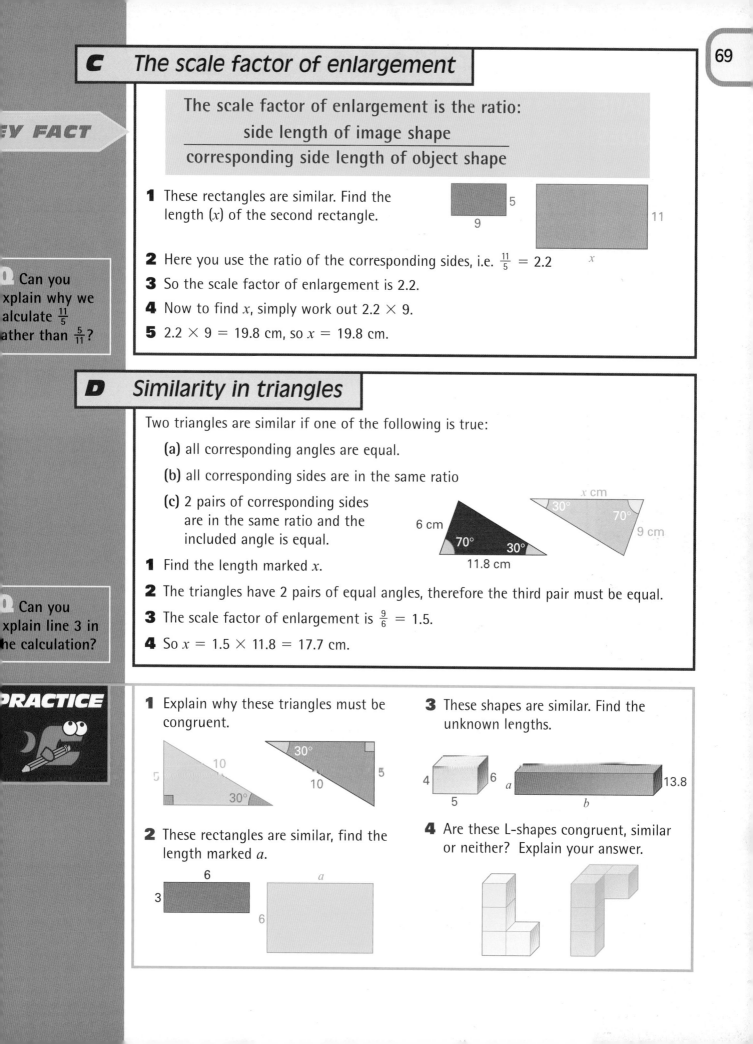

PRACTICE

1 Explain why these triangles must be congruent.

2 These rectangles are similar, find the length marked a.

3 These shapes are similar. Find the unknown lengths.

4 Are these L-shapes congruent, similar or neither? Explain your answer.

THE BARE BONES
➤ The area of any triangle can be calculated using trigonometry.
➤ The lengths of two sides and the angle between them are all that is required to find the area.

A The area of a triangle

1 This diagram shows a general triangle, which is labelled ABC.

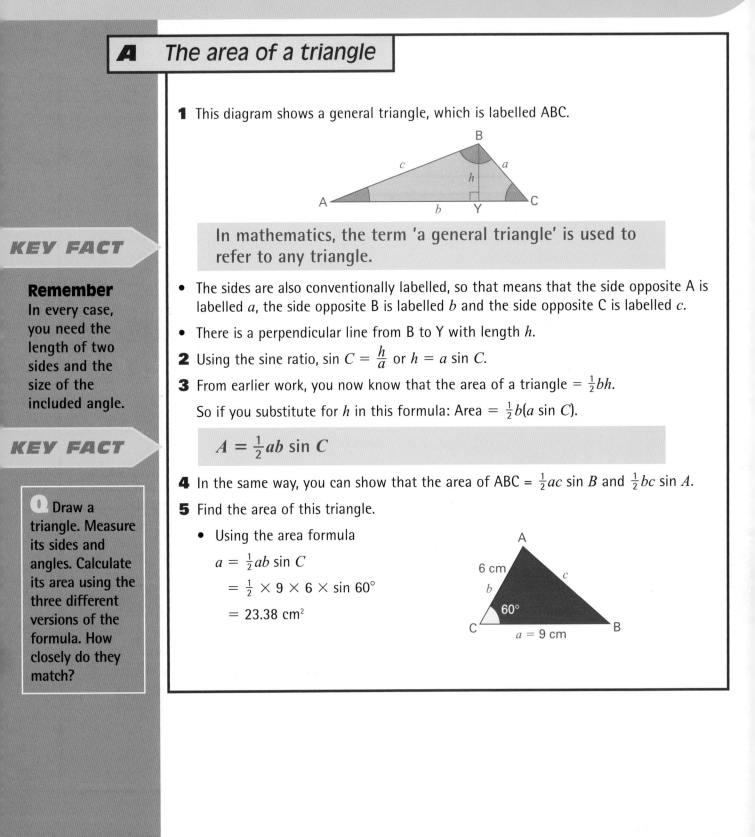

In mathematics, the term 'a general triangle' is used to refer to any triangle.

KEY FACT

Remember
In every case, you need the length of two sides and the size of the included angle.

• The sides are also conventionally labelled, so that means that the side opposite A is labelled a, the side opposite B is labelled b and the side opposite C is labelled c.

• There is a perpendicular line from B to Y with length h.

2 Using the sine ratio, $\sin C = \frac{h}{a}$ or $h = a \sin C$.

3 From earlier work, you now know that the area of a triangle $= \frac{1}{2}bh$.

So if you substitute for h in this formula: Area $= \frac{1}{2}b(a \sin C)$.

KEY FACT

$$A = \frac{1}{2}ab \sin C$$

4 In the same way, you can show that the area of ABC $= \frac{1}{2}ac \sin B$ and $\frac{1}{2}bc \sin A$.

5 Find the area of this triangle.

• Using the area formula

$a = \frac{1}{2}ab \sin C$

$= \frac{1}{2} \times 9 \times 6 \times \sin 60°$

$= 23.38 \text{ cm}^2$

Q Draw a triangle. Measure its sides and angles. Calculate its area using the three different versions of the formula. How closely do they match?

B Using Area = $\frac{1}{2}ab\sin C$ to solve problems

In order to test your understanding of the use of the formula, examiners may ask you to solve a problem using it.

1 A farmer fences off a triangular plot of land to make a pony paddock for her daughter. The paddock is an isosceles triangle with one side of 78m and two equal sides of 45m.

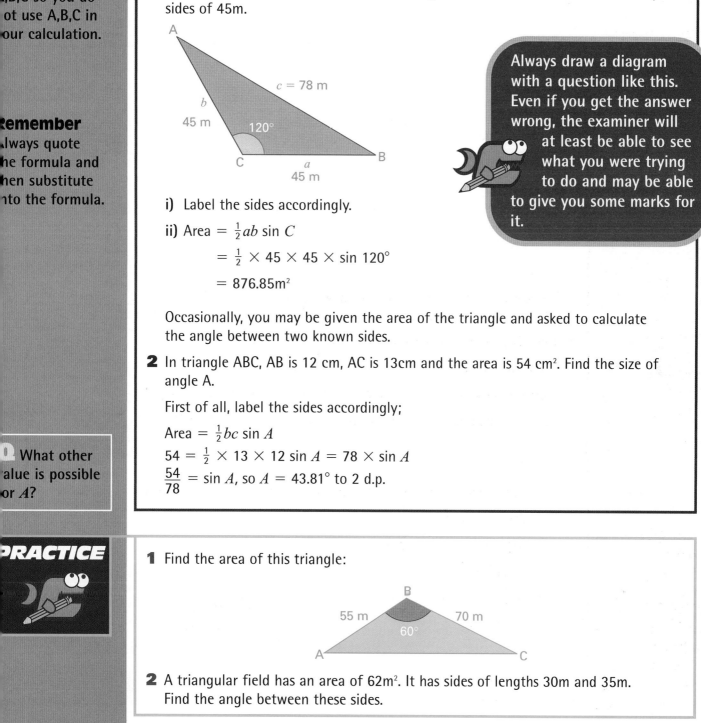

Remember
Here the sides are not listed as A,B,C so you do not use A,B,C in your calculation.

Remember
Always quote the formula and then substitute into the formula.

Always draw a diagram with a question like this. Even if you get the answer wrong, the examiner will at least be able to see what you were trying to do and may be able to give you some marks for it.

i) Label the sides accordingly.

ii) Area = $\frac{1}{2}ab\sin C$

$= \frac{1}{2} \times 45 \times 45 \times \sin 120°$

$= 876.85m^2$

Occasionally, you may be given the area of the triangle and asked to calculate the angle between two known sides.

2 In triangle ABC, AB is 12 cm, AC is 13cm and the area is 54 cm². Find the size of angle A.

First of all, label the sides accordingly;

Area = $\frac{1}{2}bc\sin A$

$54 = \frac{1}{2} \times 13 \times 12 \sin A = 78 \times \sin A$

$\frac{54}{78} = \sin A$, so $A = 43.81°$ to 2 d.p.

Q What other value is possible for A?

PRACTICE

1 Find the area of this triangle:

2 A triangular field has an area of 62m². It has sides of lengths 30m and 35m. Find the angle between these sides.

The sine rule

➤ There are trigonometric links between the sides and angles in all triangles, not just right-angled ones.

➤ The sine rule links the length of each side to the sine of the angle opposite it in the triangle.

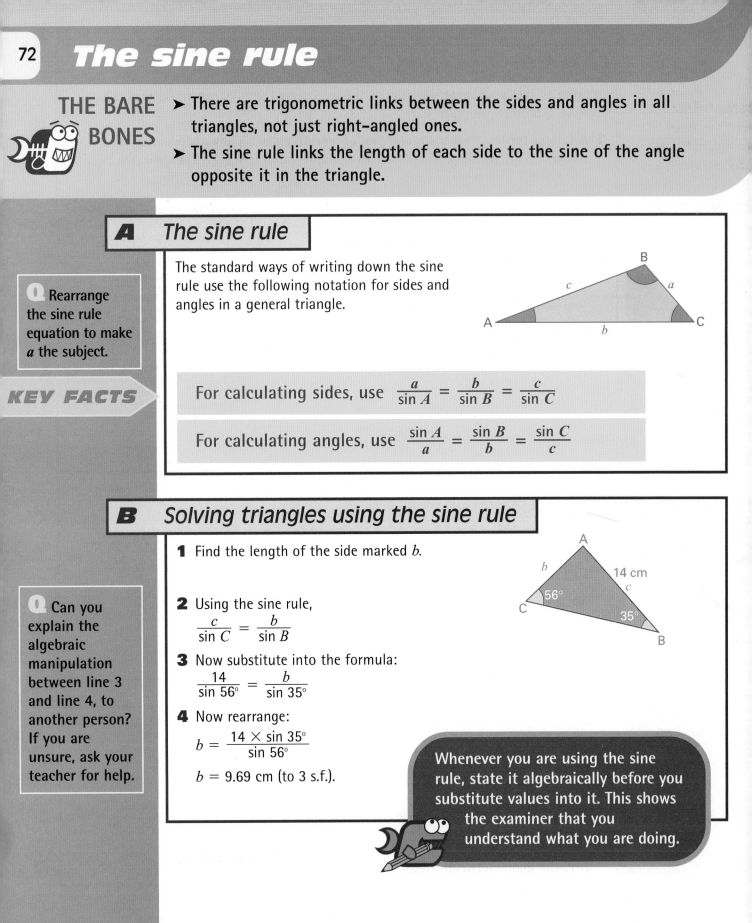

A The sine rule

Q Rearrange the sine rule equation to make *a* the subject.

The standard ways of writing down the sine rule use the following notation for sides and angles in a general triangle.

KEY FACTS

For calculating sides, use $\dfrac{a}{\sin A} = \dfrac{b}{\sin B} = \dfrac{c}{\sin C}$

For calculating angles, use $\dfrac{\sin A}{a} = \dfrac{\sin B}{b} = \dfrac{\sin C}{c}$

B Solving triangles using the sine rule

1 Find the length of the side marked *b*.

Q Can you explain the algebraic manipulation between line 3 and line 4, to another person? If you are unsure, ask your teacher for help.

2 Using the sine rule,

$$\frac{c}{\sin C} = \frac{b}{\sin B}$$

3 Now substitute into the formula:

$$\frac{14}{\sin 56°} = \frac{b}{\sin 35°}$$

4 Now rearrange:

$$b = \frac{14 \times \sin 35°}{\sin 56°}$$

$$b = 9.69 \text{ cm (to 3 s.f.)}.$$

Whenever you are using the sine rule, state it algebraically before you substitute values into it. This shows the examiner that you understand what you are doing.

C The ambiguous case

Given two sides and the angle between them, only one triangle is possible. When the angle is not the included angle two different angles are possible.

1 In triangle ABC, AB = 8, BC = 5, and $A = 29°$.

2 The triangle has been drawn for you BUT there are in fact two possible values for the size of the angle.

3 A pair of compasses has been placed at B and an arc drawn of radius 5 cm, which cuts AC at both C and c'.

This shows that there are two possible values for angle C: one is the acute angle at C, whilst the other is the obtuse angle at c'.

4 Using the sine rule:

$$\frac{\sin C}{c} = \frac{\sin A}{a}$$

$$\frac{\sin C}{8} = \frac{\sin 29°}{5}$$

$$\sin C = \frac{8 \times \sin 29°}{5} \qquad C = 50.87°$$

BUT since $\sin(180° - x) = \sin x$, you could also have $C = 180° - 50.87° = 129.13°$.

Remember
You only need to use two out of three ratios in the sine rule.

Q Can you draw a sine curve and use it to explain how two different angles can have the same sine?

PRACTICE

Remember
The sine rule can be used to find unknown sides or unknown angles.

1 Find the length of the side marked AB.

2 Find the length of WY.

Diagrams not to scale

3 Find the side marked x.

4 Find the side marked y.

5 Find the angle marked c.

6 Find the angle marked z.

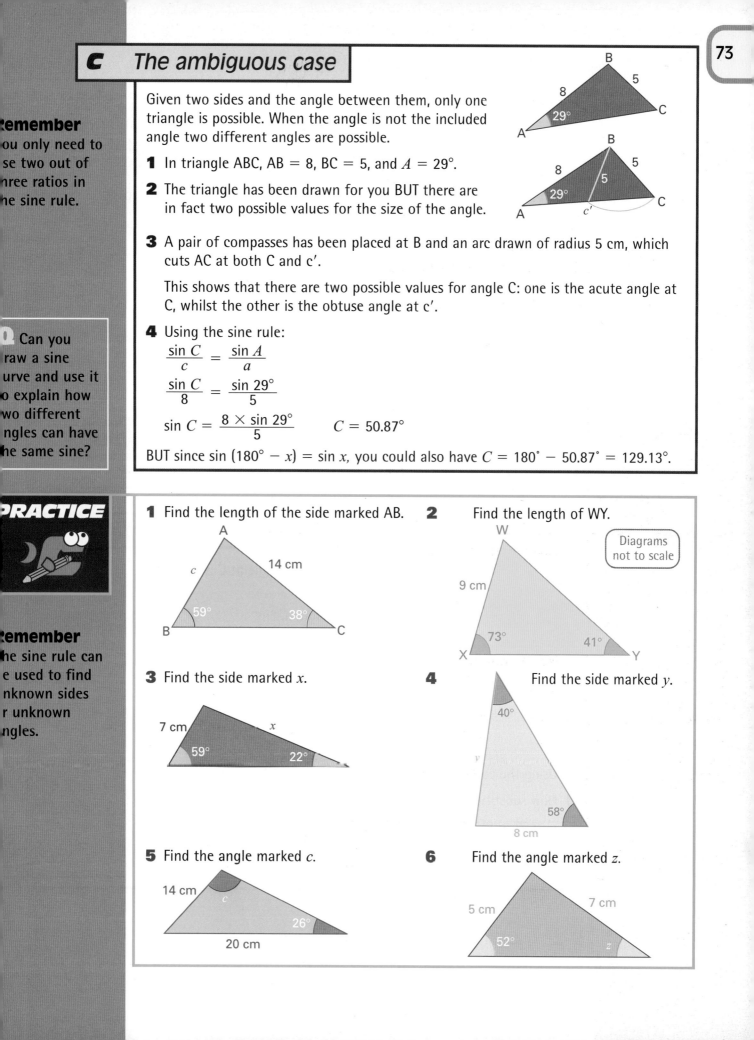

The cosine rule

THE BARE BONES

➤ Pythagoras' rule describes the link between the squares of the sides in a right-angled triangle. The cosine rule does the same in any triangle by introducing a new term involving the cosine of the angle opposite the side being calculated.

➤ The sine and cosine rules can be used in combination to solve triangles quickly and efficiently.

A The cosine rule

KEY FACT

Using the standard notation, the cosine rule states that:

$$a^2 = b^2 + c^2 - 2bc \cos A.$$

As the vertices of the triangle can be named A, B and C any way you please, there are two other versions of the cosine rule:

$$b^2 = c^2 + a^2 - 2ca \cos B.$$

$$c^2 = a^2 + b^2 - 2ab \cos C.$$

Q What happens if angle A is a right angle?

B Finding the length of a side using the cosine rule

1 In triangle ABC, AC = 9 cm, BC = 7 cm, the angle at C is 42°. Find the length of AB.

2 As always, if they do not give you a diagram, make sure you draw one. Here is an example diagram:

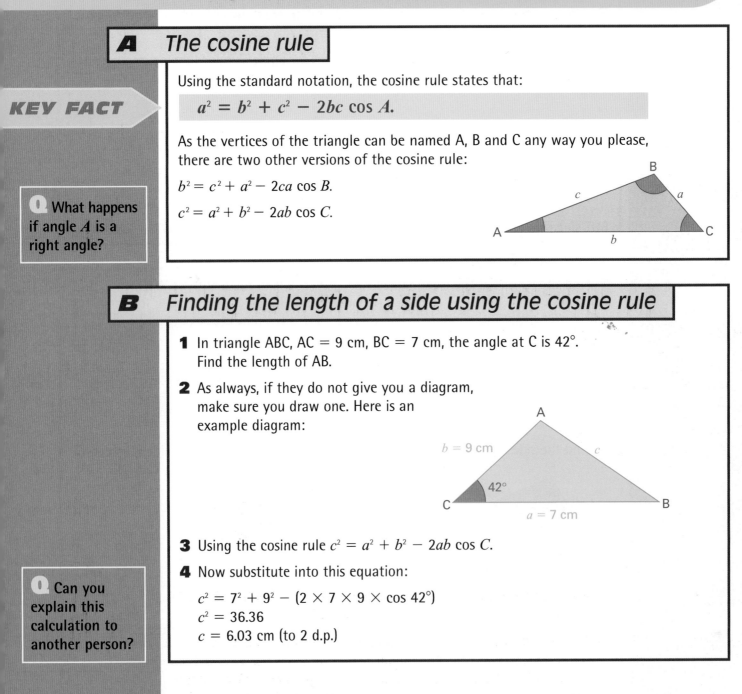

3 Using the cosine rule $c^2 = a^2 + b^2 - 2ab \cos C$.

4 Now substitute into this equation:

$$c^2 = 7^2 + 9^2 - (2 \times 7 \times 9 \times \cos 42°)$$
$$c^2 = 36.36$$
$$c = 6.03 \text{ cm (to 2 d.p.)}$$

Q Can you explain this calculation to another person?

C Finding an unknown angle using the cosine rule

When calculating an angle, rearrange $a^2 = b^2 + c^2 - 2bc \cos A$, to read $\cos A = \dfrac{b^2 + c^2 - a^2}{2bc}$

AB = 4 cm, BC = 6cm, AC = 8cm. Find the angle at A.

1 Using the cosine rule:

$$\cos A = \frac{b^2 + c^2 - a^2}{2bc}$$

$$= \frac{8^2 + 4^2 - 6^2}{2 \times 8 \times 4}$$

$\cos A = 0.6875$

Therefore $\Rightarrow A = 46.58°$ (2d.p.)

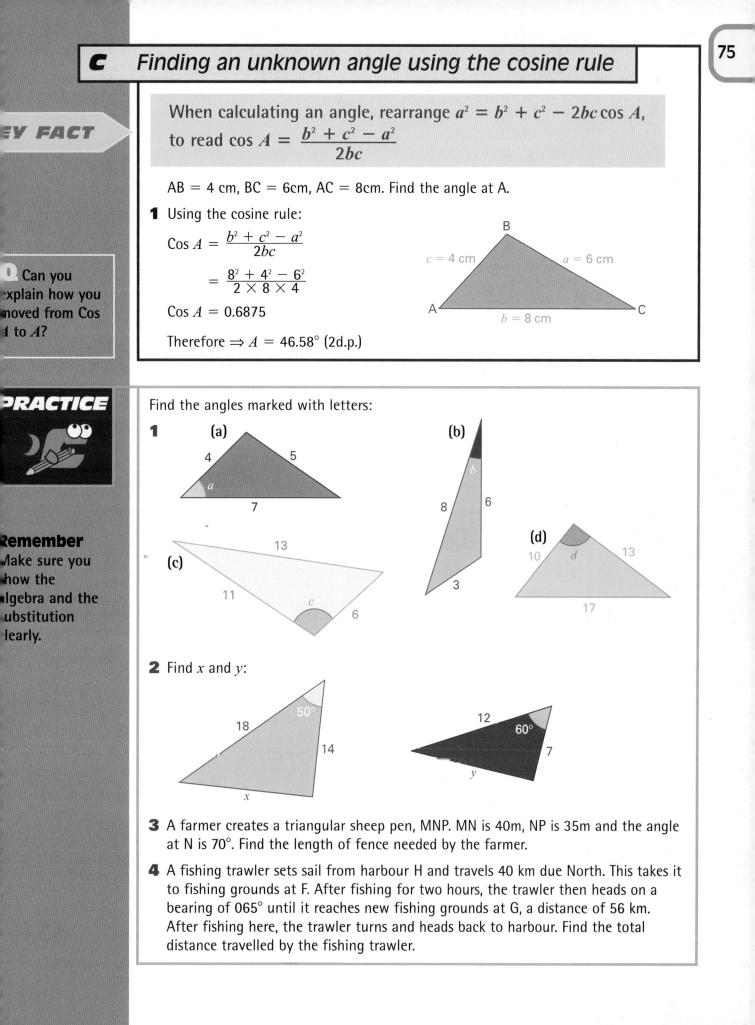

Q Can you explain how you moved from Cos A to A?

PRACTICE

Find the angles marked with letters:

1

(a)

4, 5, 7, a

(b)

8, 6, 3, b

(c)

13, 11, 6, c

(d)

10, d, 13, 17

Remember Make sure you show how the algebra and the substitution clearly.

2 Find x and y:

18, 50°, 14, x

12, 60°, 7, y

3 A farmer creates a triangular sheep pen, MNP. MN is 40m, NP is 35m and the angle at N is 70°. Find the length of fence needed by the farmer.

4 A fishing trawler sets sail from harbour H and travels 40 km due North. This takes it to fishing grounds at F. After fishing for two hours, the trawler then heads on a bearing of 065° until it reaches new fishing grounds at G, a distance of 56 km. After fishing here, the trawler turns and heads back to harbour. Find the total distance travelled by the fishing trawler.

THE BARE BONES

➤ Finding the space diagonal of a cuboid uses Pythagoras' rule in three dimensions.
➤ Trigonometry can be used in three-dimensional situations.
➤ It is usually best to split 3-D problems up into a number of 2-D diagrams.

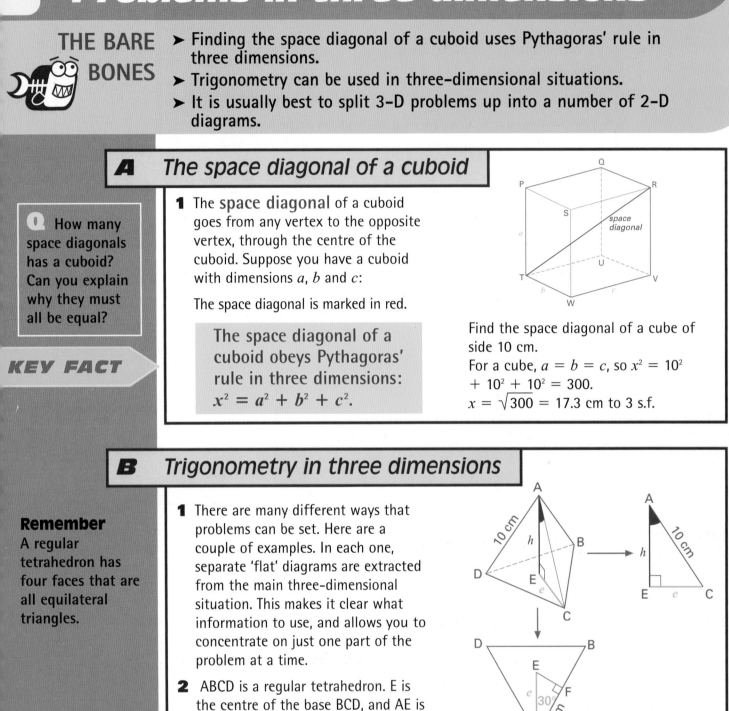

A The space diagonal of a cuboid

Q How many space diagonals has a cuboid? Can you explain why they must all be equal?

KEY FACT

1 The **space diagonal** of a cuboid goes from any vertex to the opposite vertex, through the centre of the cuboid. Suppose you have a cuboid with dimensions a, b and c:

The space diagonal is marked in red.

> The space diagonal of a cuboid obeys Pythagoras' rule in three dimensions: $x^2 = a^2 + b^2 + c^2$.

Find the space diagonal of a cube of side 10 cm.
For a cube, $a = b = c$, so $x^2 = 10^2 + 10^2 + 10^2 = 300$.
$x = \sqrt{300} = 17.3$ cm to 3 s.f.

B Trigonometry in three dimensions

Remember
A regular tetrahedron has four faces that are all equilateral triangles.

1 There are many different ways that problems can be set. Here are a couple of examples. In each one, separate 'flat' diagrams are extracted from the main three-dimensional situation. This makes it clear what information to use, and allows you to concentrate on just one part of the problem at a time.

2 ABCD is a regular tetrahedron. E is the centre of the base BCD, and AE is perpendicular to the base. Calculate angle EAC.

3 The final answer will come from triangle EAC (on the right of the main diagram), but to be able to use this, you need to know the length of EC (marked e in the diagram). e can be calculated from the base, triangle BCD (below main diagram).

The midpoint of BC has been labelled F.

In triangle CEF, $\cos 30° = \dfrac{\text{adjacent}}{\text{hypotenuse}} = \dfrac{5}{e}$

4 As $\cos 30° = \dfrac{\sqrt{3}}{2}$, $e = \dfrac{10}{\sqrt{3}}$. Leave this in surd form for the moment.

In triangle EAC, $\sin \text{EAC} = \dfrac{e}{10} = \dfrac{1}{\sqrt{3}}$. So angle $\text{EAC} = \sin^{-1}\left(\dfrac{1}{\sqrt{3}}\right) = 35.3°$ to 1 d.p.

5 In the cuboid shown, solve triangle ACH.

First, find the lengths (a, b, c) of the sides, using Pythagoras' rule on the faces of the cuboid.

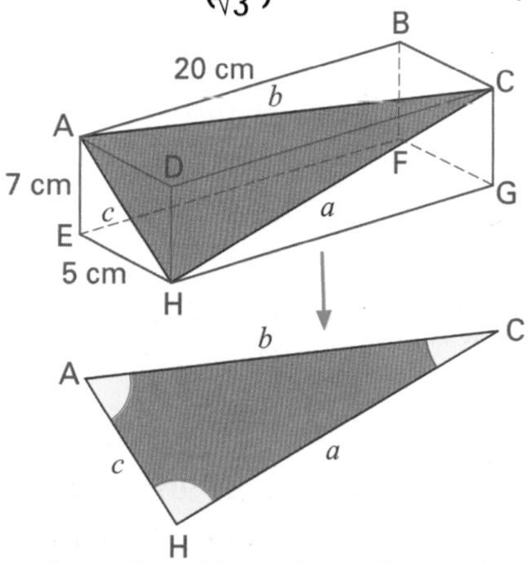

This gives $a = \sqrt{449} = 21.2$ cm to 1 d.p.,

$\qquad b = \sqrt{425} = 20.6$ cm to 1 d.p.,

$\qquad c = \sqrt{74} = 8.6$ cm to 1 d.p.

6 Using the diagram above, the cosine rule states that $a^2 = b^2 + c^2 - 2bc\cos A$,

so that $A = \cos^{-1}\left(\dfrac{b^2 + c^2 - a^2}{2bc}\right) = \cos^{-1}\left(\dfrac{425 + 74 - 449}{2 \times \sqrt{425 \times 74}}\right) = 81.9°$ to 1 d.p.

The sine rule then states that $\dfrac{\sin C}{c} = \dfrac{\sin A}{a}$, so that

$C = \sin^{-1}\left(\dfrac{c \sin A}{a}\right) = \sin^{-1}\left(\dfrac{449 \sin 81.895...°}{\sqrt{74}}\right) = 23.7°$ to 1 d.p.

$\qquad H = 180° - A - C = 74.4°$ to 1 d.p.

All the sides and angles have now been found.

Remember

Draw a 'flat' diagram of the triangle you're interested in.

Remember

Retain accurate values in the calculator memory for the next part of the question.

Q Can you explain how angle H is calculated?

PRACTICE

1 Calculate the space diagonal of:
   (a) a cuboid 2 cm by 3 cm by 4 cm.
   (b) a cuboid 10 cm by 5 cm by 20 cm.
   (c) a unit cube.

2 John made a square-based pyramid out of wire. The perpendicular height of his model was 20 cm and the base was 10 cm on a side. Assuming that he only constructed the 8 edges of the pyramid, what length of wire did John use?

3 The diagram shows two equally tall observers, A and B, looking at a transmitter mast CM which is 10 m taller than they are. A is due south of C, and the angle of elevation of M from her position is 16°. B is due east of C and registers an angle of elevation of 32°. What is the bearing of B from A?

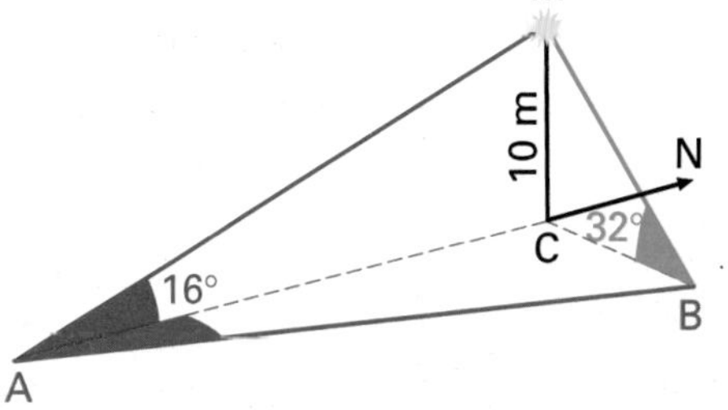

Transformations

THE BARE BONES
> A transformation is an operation that can change the position, shape and size of an object.
> In mathematical terms, a transformation maps an object to its image.

A Reflections

1 An object reflected in a mirror creates an image.

2 The object and the image are symmetrical about the mirror line.

3 The mirror line is an axis of symmetry.

KEY FACT

The object and the image are equidistant from the mirror line.

Q What happens to your reflection when you raise your left hand?

B Translation

KEY FACT

A translation is 'sliding' movement.

1 The image of a shape after a translation is congruent to its object, and not rotated or reflected. In the diagram, object triangle A is translated to B by moving 5 units 'right' (parallel to the positive x-axis) and 4 units 'down' (parallel to the negative y-axis).

2 Translations are usually described using vectors. The vector describing '5 right and 4 units down' is written $\begin{pmatrix} 5 \\ -4 \end{pmatrix}$. 'Right' and 'up' are given positive values, 'left' and 'down', negative ones.

Q Can you write down the vector for the shift from (2, 3) to (−3, 4)?

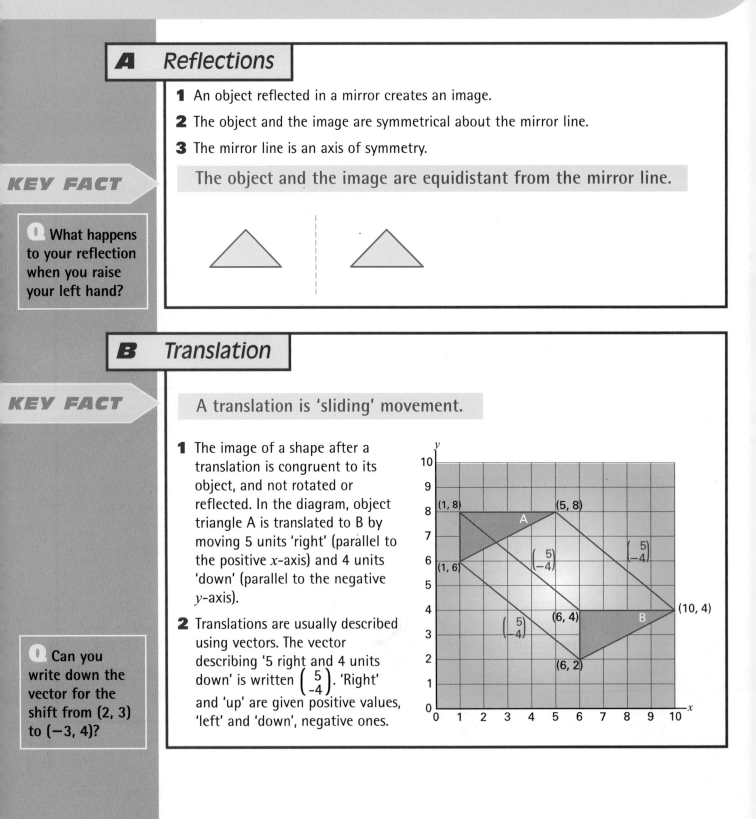

c Rotations

1 A shape can be rotated about a point called the centre of rotation (marked by a cross on the diagram).

2 You need to specify an angle and direction of rotation (clockwise or anti-clockwise, unless the angle is 180°).

3 To find the centre of rotation:

(a) join two pairs of corresponding points on the object and image with straight lines.

(b) draw in the perpendicular bisectors of these lines. The location where they meet is the centre of rotation.

Q Can you find the centre of rotation using this method?

PRACTICE

Remember
The centre of rotation is often the origin at (0, 0).

1 Reflect this triangle in the line $y = x$.

2 Rotate this rectangle by 180° about (0, 2).

3 Translate this rectangle using the vector $\binom{3}{4}$.

Enlargements

THE BARE BONES

➤ An enlargement occurs when the size of a shape is changed by a transformation.

➤ An enlargement is specified by its centre of enlargement and its scale factor.

A Enlargements

1 When you have photographs developed, different-sized prints can be made from the same negative.

2 The length and width of a print are in the same ratio as the length and width of the negative.

3 All lengths on the image are increased or reduced according to the scale factor of enlargement.

4 Copy and enlarge the triangle XYZ, using the point A as the centre of enlargement, with scale factor 2 (label it X'Y'Z').
Make XX' = AX, YY' = AY and ZZ' = AZ.

KEY FACT

Enlargements are similar to their objects.

Q Can you draw the enlargement that would occur with a scale factor of 4?

5 Each side of X'Y'Z' needs to be twice as long as the corresponding sides on XYZ.

6 The corresponding angles in both triangles are the same.

7 Each side in the enlarged shape is parallel to the relevant corresponding side in the original shape.

B Enlarging by a fractional scale factor less than one

KEY FACT

An enlargement with a scale factor less than 1 will actually reduce the size of the shape.

Q Redraw the triangle and see if you can enlarge it using the origin as the centre and a scale factor of enlargement of $\frac{1}{3}$.

1 Here, enlarging by a scale factor of less than one actually means reducing the size of the object.

2 Enlarge triangle ABC with a scale factor of $\frac{1}{2}$ using the origin as the centre of enlargement.

3 Notice here how every line on the image is exactly $\frac{1}{2}$ the length of the line on the object.

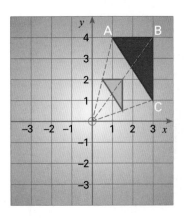

c Enlarging with a negative scale factor

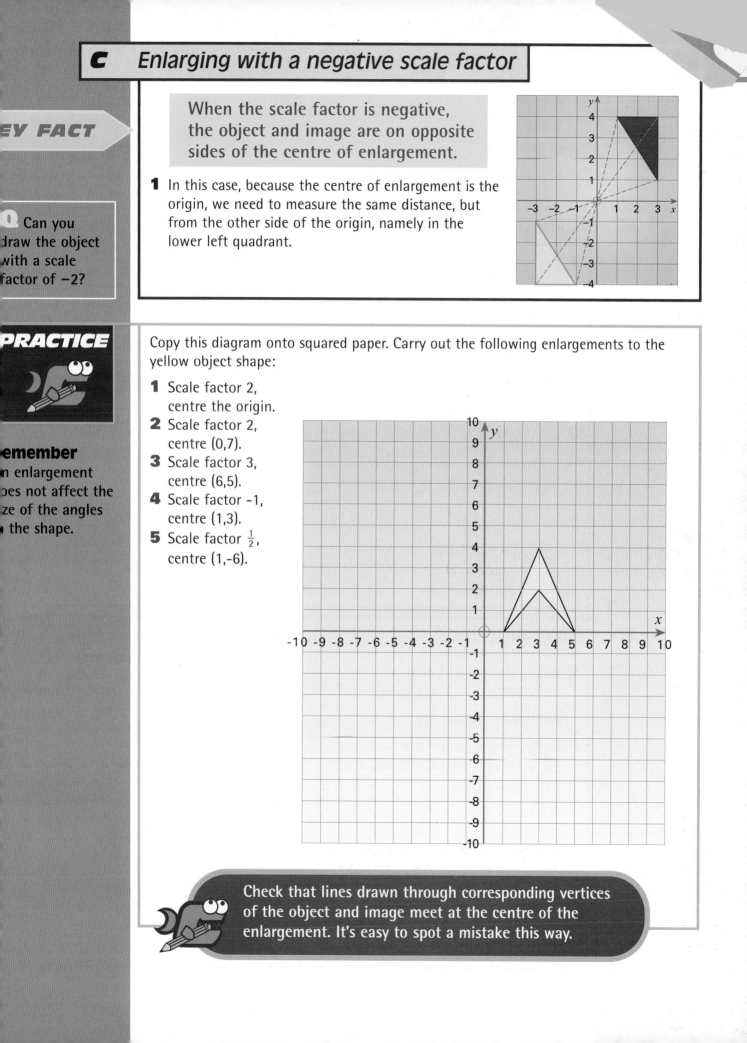

EY FACT

When the scale factor is negative, the object and image are on opposite sides of the centre of enlargement.

1 In this case, because the centre of enlargement is the origin, we need to measure the same distance, but from the other side of the origin, namely in the lower left quadrant.

Q Can you draw the object with a scale factor of −2?

PRACTICE

emember
n enlargement
oes not affect the
ze of the angles
the shape.

Copy this diagram onto squared paper. Carry out the following enlargements to the yellow object shape:

1 Scale factor 2, centre the origin.

2 Scale factor 2, centre (0,7).

3 Scale factor 3, centre (6,5).

4 Scale factor −1, centre (1,3).

5 Scale factor $\frac{1}{2}$, centre (1,−6).

Check that lines drawn through corresponding vertices of the object and image meet at the centre of the enlargement. It's easy to spot a mistake this way.

Functions

THE BARE BONES

➤ A function is defined as a rule which operates on a number and turns it into another number.

➤ Many functions have opposites or inverses. The inverse function of f is written f^{-1}.

A *Introduction to function notation*

1 You can think of a function as a number machine that changes an input x into an output $f(x)$.

$$x \longrightarrow \boxed{f} \longrightarrow f(x)$$

The output of $f(x)$ is read as 'f of x'.

> **KEY FACT**
>
> If a function f is 'cube' then, for an input x, the output is x^3.

2 In this case, for an input of 2, the output will be 8. Mathematically, this is written as $f(2) = 2^3 = 8$.

3 In the same way, $f(-3) = -27$, and $f(m + 2) = (m + 2)^3 = m^3 + 6m^2 + 12m + 8$.

> **KEY FACT**
>
> A function can also be the same as a sequence of operations or rules.

Q Can you rewrite $f(2m)$ in the same manner as line 3?

4 For instance, $g(x) = 3x^2 + 2$ is the output from:

$$\text{Input} \longrightarrow \boxed{\text{Square}} \longrightarrow \boxed{\times 3} \longrightarrow \boxed{+ 2} \longrightarrow 3x^2 + 2$$

B *Using functions*

1 When $f(x) = x^2 + 3$, find:

(a) $f(2)$ (b) $f(-5)$ (c) $f(1)$

(d) $f(0)$ (e) $f(-t)$ (f) $f(t + 1)$

2 (a) replacing x by 2, $f(2) = 2^2 + 3 = 7$.

(b) replacing x by -5, $f(-5) = (-5)^2 + 3 = 28$.

(c) replacing x by 1, $f(1) = 1^2 + 3 = 4$.

(d) replacing x by 0, $f(0) = 0 + 3 = 3$.

(e) replacing x by $-t$, $f(-t) = (-t)^2 + 3 = t^2 + 3$.

(f) replacing x by $(t + 1)$, $f(t+1) = (t + 1)^2 + 3 = t^2 + 2t + 4$.

Q Can you explain this example to another person?

Whenever you are unsure in maths, put numbers into the question instead of variables. Look for patterns and ask yourself, what is 'the same' about these numbers or this operation? This is called specialising and is a good tip for getting unstuck.

Functions and equations

1 Given $f(x) = 2x^2 + 1$, find the values of x, when $f(x) = 33$.

2 When $f(x) = 33$ $33 = 2x^2 + 1$

3 Solve the equation $33 = 2x^2 + 1$

$$32 = 2x^2$$
$$x^2 = 16$$
$$x = \sqrt{16}$$
$$x = \pm\,4$$

Q For x to be ± 5, what would $f(x)$ be?

D *Inverse functions*

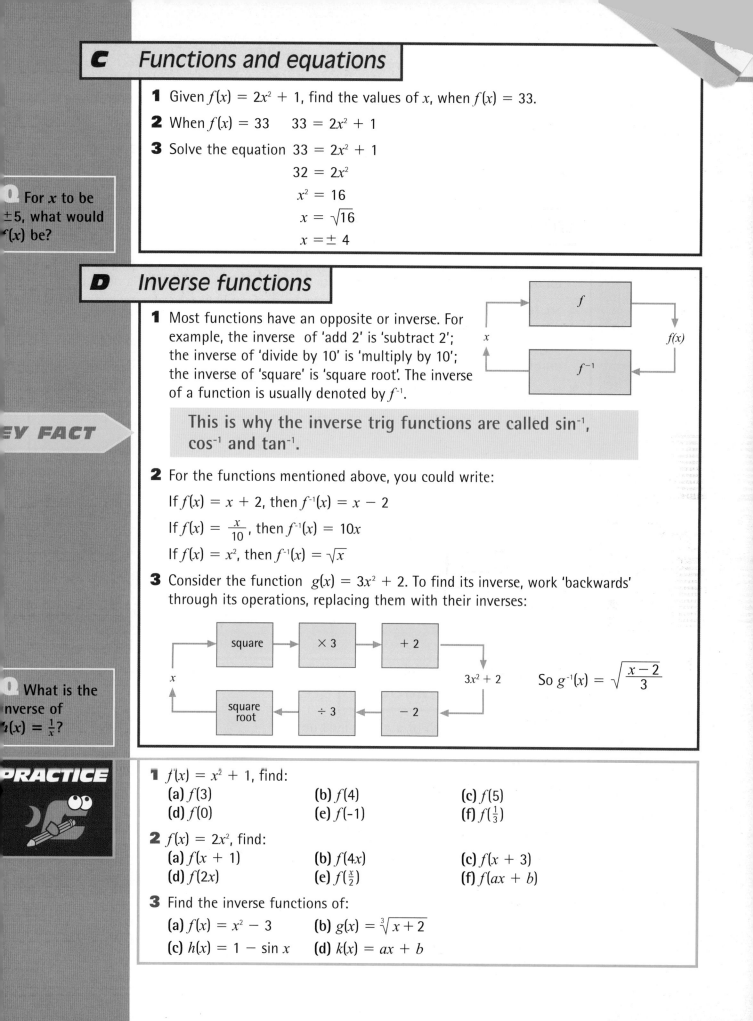

1 Most functions have an opposite or inverse. For example, the inverse of 'add 2' is 'subtract 2'; the inverse of 'divide by 10' is 'multiply by 10'; the inverse of 'square' is 'square root'. The inverse of a function is usually denoted by f^{-1}.

KEY FACT

> This is why the inverse trig functions are called $\sin^{-1}$, $\cos^{-1}$ and $\tan^{-1}$.

2 For the functions mentioned above, you could write:

If $f(x) = x + 2$, then $f^{-1}(x) = x - 2$

If $f(x) = \frac{x}{10}$, then $f^{-1}(x) = 10x$

If $f(x) = x^2$, then $f^{-1}(x) = \sqrt{x}$

3 Consider the function $g(x) = 3x^2 + 2$. To find its inverse, work 'backwards' through its operations, replacing them with their inverses:

square → × 3 → + 2

x → → → $3x^2 + 2$

square root ← ÷ 3 ← − 2

So $g^{-1}(x) = \sqrt{\dfrac{x - 2}{3}}$

Q What is the inverse of $h(x) = \frac{1}{x}$?

PRACTICE

1 $f(x) = x^2 + 1$, find:
(a) $f(3)$ (b) $f(4)$ (c) $f(5)$
(d) $f(0)$ (e) $f(-1)$ (f) $f(\frac{1}{3})$

2 $f(x) = 2x^2$, find:
(a) $f(x + 1)$ (b) $f(4x)$ (c) $f(x + 3)$
(d) $f(2x)$ (e) $f(\frac{x}{2})$ (f) $f(ax + b)$

3 Find the inverse functions of:
(a) $f(x) = x^2 - 3$ (b) $g(x) = \sqrt[3]{x + 2}$
(c) $h(x) = 1 - \sin x$ (d) $k(x) = ax + b$

Translating graphs

THE BARE BONES
➤ Vertical and horizontal translations of graphs have simple algebraic representations.
➤ Any translation can be achieved by combining horizontal and vertical components.

A 'Vertical' translations

1 Suppose you have a graph of some function $y = f(x)$. The graph can be translated up or down by taking the expression for y and adding or subtracting a number.

2 So, for example, the graph of $y = \sin x + 1$ is the same as the graph of $y = \sin x$, translated 1 unit in the positive y direction:

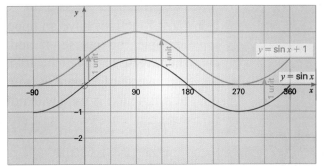

3 Here is the graph of $y = x^2$, translated 3 units in the negative y direction to become $y = x^2 - 3$:

KEY FACT

In general, to translate $y = f(x)$ by a units in the positive y direction, replace it by $y = f(x) + a$. To translate it by a units in the negative y direction, replace it by $y = f(x) - a$.

Q On one set of axes, sketch the graphs of $y = \cos x$ and $y = \cos x - 2$.

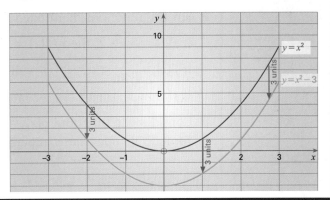

B 'Horizontal' translations

1 To translate the graph of $y = f(x)$ by 3 units in the positive x direction, replace x by $x - 3$.

2 So, for example, the graph of $y = \dfrac{1}{x - 3}$ is the same as the graph of $y = \dfrac{1}{x}$, translated 3 units in the positive x direction (see diagram).

Q On one set of axes, sketch the graphs of $y = x^2$ and $y = (x - 1)^2$.

3 The graph of $y = x^3$ translates 2 units in the negative y direction to become $y = (x + 2)^3$.

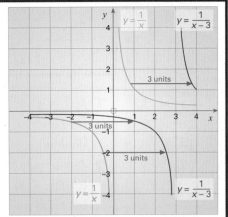

KEY FACT

In general, to translate $y = f(x)$ by a units in the positive x direction, replace it by $y = f(x - a)$. To translate it by a units in the negative x direction, replace it by $y = f(x + a)$.

C Double translations

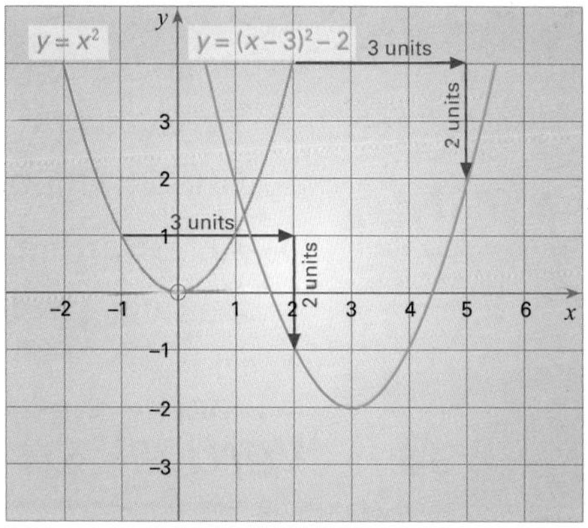

1 By combining the techniques of the last two sections, you can translate a graph in any desired way.

2 So, for example, to translate the graph of $y = x^2$ by the vector $\begin{pmatrix} 3 \\ -2 \end{pmatrix}$ follow these steps:

- Replace x by $(x - 3)$ to translate 3 units to the right: $y = (x - 3)^2$.
- Subtract 2 from y to translate 2 units down: $y = (x - 3)^2 - 2$.

3 Note that the quadratic expression $(x - 3)^2 - 2$ can be expanded and simplified to give $x^2 - 6x + 7$, so the translated graph should be that of $y = x^2 - 6x + 7$.

Q By checking suitable pairs of co-ordinates, satisfy yourself that: $y = x^2 - 6x + 7$ is the correct translation of $y = x^2$.

EY FACT

In general, to translate $y = f(x)$ by the vector $\begin{pmatrix} a \\ b \end{pmatrix}$, replace it by $y = f(x - a) + b$.

PRACTICE

In these questions, 'right' and 'left' refer to the positive and negative x directions, respectively. 'Up' and 'down' refer to the positive and negative y directions, respectively.

1 Draw the graph of $y = x^2$. Translate the graph as follows and write down the equation of the new curve.

(a) 4 units up.
(b) 3 units left.
(c) 5 units down.
(d) 7 units right.
(e) 6 units up and 2 units left.
(f) 4 units right and 1 unit down.

2 Show how $y = \frac{1}{x}$ can translate to:

$$y = \frac{3x + 7}{x + 2}$$

member Write $3x + 7$ in the form $p(x + 2) + q$, where p and q are integers.

3 (a) Explain how to produce the graph of $y = x^2 + 8x + 12$ by applying transformations to $y = x^2$.

(b) The graph of $y = x^2 + 8x + 12$ is a **parabola**. Write down the co-ordinates of its vertex.

(c) Illustrate your answer with a sketch.

Remember Complete the square for the expression $x^2 + 8x + 12$.

4 This graph shows a **trigonometric function** that has been translated. Find the equation of the function shown on the graph.

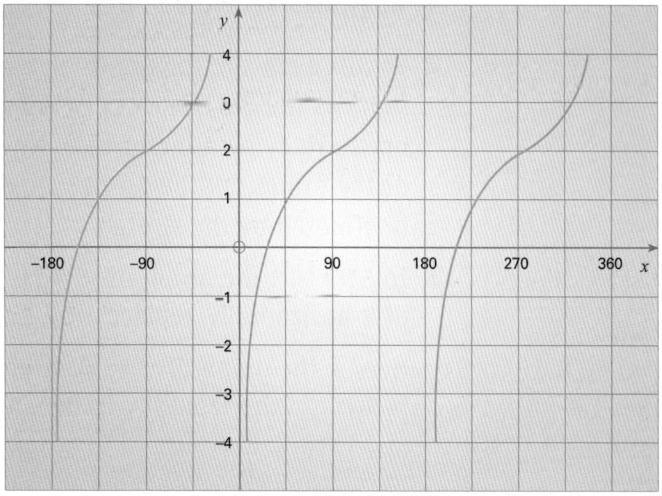

THE BARE BONES

➤ Vertical and horizontal 'stretches' and 'squashes' of graphs have simple algebraic representations.

➤ An enlargement can be achieved by combining horizontal and vertical stretches.

A 'Vertical' stretches

1 Suppose you have a graph of some function $y = f(x)$. The graph can be stretched vertically by multiplying every y value by some number a.

2 So, for example, the graph of $y = 3\cos x$ is the same as the graph of $y = \cos x$, stretched vertically 3 times. Note that points of the graph that lie on the x-axis are unaffected by the stretch.

3 Graphs are vertically 'squashed' if $a < 1$. Here is the graph of $y = x^2$, squashed to half its height to become $y = \frac{1}{2}x^2$:

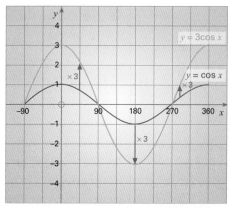

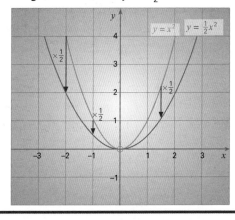

KEY FACT

In general, to stretch $y = f(x)$ vertically by a factor of a, replace it by $y = af(x)$.

Q On one set of axes, sketch the graphs of $y = \sin x$ and $y = 2\sin x$.

B 'Horizontal' stretches

1 To **squash** the graph of $y = f(x)$ by a factor of 3 horizontally, replace x by $3x$.

2 So, for example, the graph of $y = \frac{1}{3x}$ is the same as the graph of $y = \frac{1}{x}$, squashed horizontally by a factor of 3. Note that points of the graph that lie on the y-axis are unaffected by the stretch.

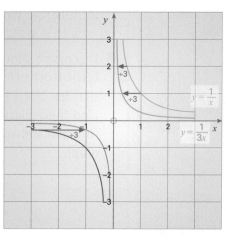

3 To **stretch** the graph of $y = f(x)$ by a factor of 2 horizontally, replace x by $\frac{1}{2}x$.

Here is the graph of $y = x^3$, stretched by a factor of 2 horizontally to become $y = (\frac{1}{2}x)^3 = \frac{x^3}{8}$:

Q On one set of axes, sketch the graphs of $y = x^2$ and $y = (2x)^2$.

B

1 Note that points of the graph that lie on the y-axis are unaffected by the stretch.

> In general, the graph of $y = f(ax)$ is the same as the graph of $y = f(x)$, horizontally squashed by a factor of a. If $a < 1$, the transformation is a stretch and increases the x scale.

KEY FACT

PRACTICE

Remember
Make sketch graphs to help, if you wish, or use software plotting.

1 For each part, write down the equation of the image graph:

	image graph	transformation
(a)	$y = x^3$	3 × vertical stretch
(b)	$y = x^2$	squash to half size vertically
(c)	$y = x^2$	2 × horizontal stretch
(d)	$y = (x + 1)^2$	squash to $\frac{1}{10}$ size horizontally
(e)	$y = x^3 + 5x^2$	5 × horizontal stretch
(f)	$y = \frac{1}{x}$	$1\frac{1}{2}$ × vertical stretch

2 For each part, describe the transformation given by the change in equation:

	object graph	image graph
(a)	$y = x^2$	$y = 5x^2$
(b)	$y = x^2$	$y = \left(\frac{x}{3}\right)^2$
(c)	$y = x^2 + 2x$	$y = 4x^2 + 8x$
(d)	$y = x^2 + 2x$	$y = 4x^2 + 4x$
(e)	$y = \tan x$	$y = \tan \frac{x}{2}$
(f)	$y = \frac{1}{x}$	$y = \frac{10}{x}$

THE BARE BONES

➤ In a column vector $\begin{pmatrix} a \\ b \end{pmatrix}$, a and b are called components of the vector.

➤ Vectors can also be written as bold single letters like x or y. When they are handwritten, they are always written with a line underneath, e.g. m or n.

A   Vectors and vector addition

Remember
The x component of the vector is the top number in the bracket.

1 Column vectors, e.g. $\begin{pmatrix} 3 \\ 2 \end{pmatrix}$ can be used to describe translations, where the x-value of the vector represents a shift parallel to the x-axis and the y-value represents a shift parallel to the y-axis. In this example, the 3 in the column vector represents a shift of 3 units to the right and the 2 represents a shift of 2 units upwards.

2 Alternatively, where A is the start point and B is the end point of a translation, the translation can be described as $\overrightarrow{AB}$. The translation that reverses $\overrightarrow{AB}$ and translates B to A is $\overrightarrow{BA}$:

$$\overrightarrow{BA} = -\overrightarrow{AB}$$

KEY FACTS

By definition, a vector has unique length and direction.

3 The vectors $\overrightarrow{WX}$ and $\overrightarrow{YZ}$ are responsible for the shift of the square from A to B and then on to C.

4 The vectors can be written as column vectors, namely $\begin{pmatrix} 3 \\ 2 \end{pmatrix}$ and $\begin{pmatrix} -4 \\ -5 \end{pmatrix}$

5 But in effect you could have translated directly from A to C and missed out the interim stage of moving the square to B altogether.

6 So this must mean that $\overrightarrow{AC}$ is the result of adding $\overrightarrow{AB}$ and $\overrightarrow{BC}$.

7 $\overrightarrow{AC}$ must therefore be the result of two successive translations, which must be:

$$\begin{pmatrix} 3 \\ 2 \end{pmatrix} + \begin{pmatrix} -4 \\ -5 \end{pmatrix} = \begin{pmatrix} -1 \\ -3 \end{pmatrix}$$

$$\overrightarrow{AB} + \overrightarrow{BC} = \overrightarrow{AC}$$

KEY FACT

You can add any two vectors by creating a triangle.

A

Q Can you explain the triangle law of addition to another person?

8 This shows the triangle law of addition. To take the route directly from X to Y, is equivalent to travelling via Z, hence you can legitimately represent $\overrightarrow{XY}$ by $\mathbf{a} + \mathbf{b}$

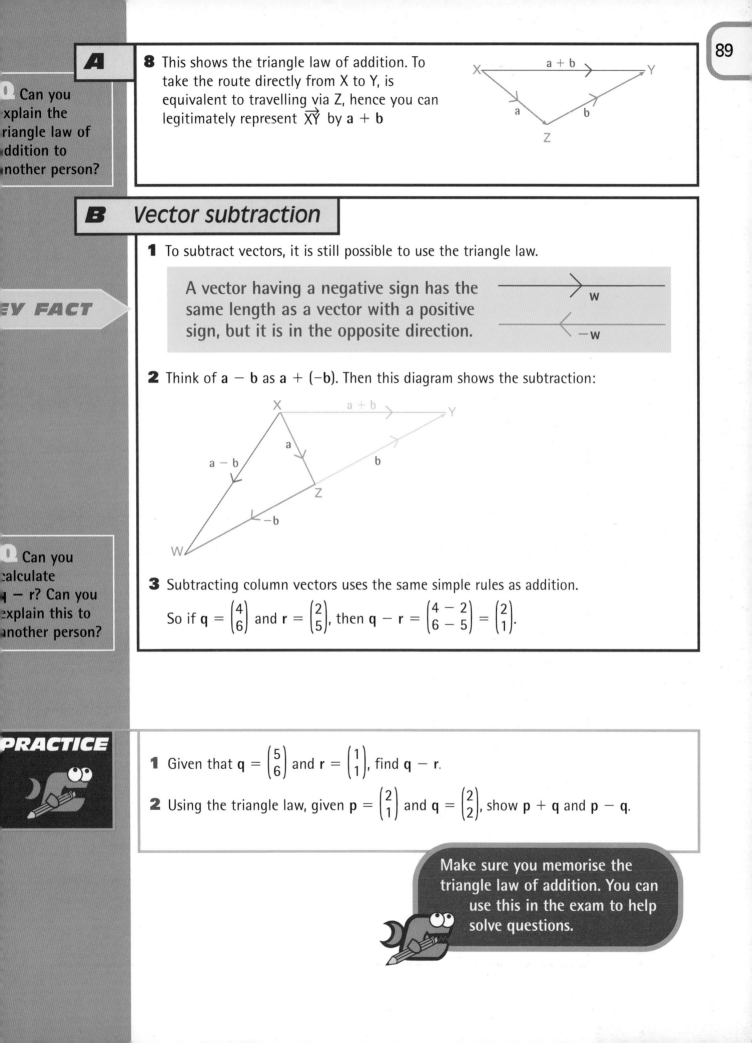

B *Vector subtraction*

1 To subtract vectors, it is still possible to use the triangle law.

EY FACT

> A vector having a negative sign has the same length as a vector with a positive sign, but it is in the opposite direction.

2 Think of $\mathbf{a} - \mathbf{b}$ as $\mathbf{a} + (-\mathbf{b})$. Then this diagram shows the subtraction:

Q Can you calculate q − r? Can you explain this to another person?

3 Subtracting column vectors uses the same simple rules as addition.

So if $\mathbf{q} = \begin{pmatrix} 4 \\ 6 \end{pmatrix}$ and $\mathbf{r} = \begin{pmatrix} 2 \\ 5 \end{pmatrix}$, then $\mathbf{q} - \mathbf{r} = \begin{pmatrix} 4 - 2 \\ 6 - 5 \end{pmatrix} = \begin{pmatrix} 2 \\ 1 \end{pmatrix}$.

PRACTICE

1 Given that $\mathbf{q} = \begin{pmatrix} 5 \\ 6 \end{pmatrix}$ and $\mathbf{r} = \begin{pmatrix} 1 \\ 1 \end{pmatrix}$, find $\mathbf{q} - \mathbf{r}$.

2 Using the triangle law, given $\mathbf{p} = \begin{pmatrix} 2 \\ 1 \end{pmatrix}$ and $\mathbf{q} = \begin{pmatrix} 2 \\ 2 \end{pmatrix}$, show $\mathbf{p} + \mathbf{q}$ and $\mathbf{p} - \mathbf{q}$.

> Make sure you memorise the triangle law of addition. You can use this in the exam to help solve questions.

Vectors 2

THE BARE BONES
➤ Vectors cannot be multiplied or divided by each other, but can be multiplied by numbers called scalars.
➤ The normal rules of algebra apply to vectors and scalars.
➤ The magnitude (length) of a vector can be determined using Pythagoras' rule.

A Multiplying vectors by scalars

1 Suppose vector **c** is $\begin{pmatrix} 3 \\ 4 \end{pmatrix}$. Then $\mathbf{c} + \mathbf{c} = \begin{pmatrix} 3 \\ 4 \end{pmatrix} + \begin{pmatrix} 3 \\ 4 \end{pmatrix} = \begin{pmatrix} 6 \\ 8 \end{pmatrix}$.

This vector can also be written as 2**c**. 2 is an ordinary number with size but no direction. These numbers are sometimes called **scalars**, to distinguish them from vectors, that have a direction.

KEY FACT

> When a vector is multiplied by a scalar, its length changes but its direction remains the same, unless the scalar is negative, when the direction is reversed.

Q If $\mathbf{v} = \begin{pmatrix} 3 \\ 4 \end{pmatrix}$, what is $\frac{1}{2}\mathbf{v}$?

2 To multiply a vector in column form by a scalar, simply multiply both components.

$$2\mathbf{c} = \begin{pmatrix} 2 \times 3 \\ 2 \times 4 \end{pmatrix} = \begin{pmatrix} 6 \\ 8 \end{pmatrix}$$

B Vector algebra

1 The normal rules of algebra apply to vectors and scalars, so you can simplify expressions containing them.

In the diagram, $\overrightarrow{AB} = \mathbf{p} + 3\mathbf{q} + 2\mathbf{p} + \mathbf{q} - 4\mathbf{p}$. Collecting like terms, this simplifies to $\overrightarrow{AB} = -\mathbf{p} + 4\mathbf{q}$ or $4\mathbf{q} - \mathbf{p}$.

KEY FACT

> $\overrightarrow{AB}$ is called the resultant of the vectors making it up.

2 Suppose that in the diagram, $\mathbf{p} = \begin{pmatrix} -2 \\ -1 \end{pmatrix}$ and $\mathbf{q} = \begin{pmatrix} 4 \\ 0 \end{pmatrix}$.

Then $\overrightarrow{AB} = 4\mathbf{q} - \mathbf{p} = \begin{pmatrix} 4 \times 4 \\ 4 \times 0 \end{pmatrix} - \begin{pmatrix} -2 \\ -1 \end{pmatrix} = \begin{pmatrix} 16 + 2 \\ 0 + 1 \end{pmatrix} = \begin{pmatrix} 18 \\ 1 \end{pmatrix}$.

Q Simplify $3\mathbf{p} - 2\mathbf{q} + \mathbf{p} + \mathbf{q}$ and write the resultant as a column vector.

3 Here is a summary of the algebraic rules for vectors and scalars. For vectors **a** and **b** and a scalar k: $\mathbf{a} - \mathbf{b} = \mathbf{a} + (-\mathbf{b})$

$$k\mathbf{a} = \mathbf{a} + \mathbf{a} + \dots + \mathbf{a} + \mathbf{a} \ (k \text{ times})$$

$$k(\mathbf{a} + \mathbf{b}) = k\mathbf{a} + k\mathbf{b}$$

C   Using position vectors

There are two particular uses of position vectors that you need to be aware of.

1   If points A and B have position vectors **a** and **b**, respectively, then $\vec{AB} = \mathbf{b} - \mathbf{a}$.

If A is the point (4, 8) and B is (6, 3), then $\vec{AB} = \mathbf{b} - \mathbf{a} = \begin{pmatrix} 6 \\ 3 \end{pmatrix} - \begin{pmatrix} 4 \\ 8 \end{pmatrix} = \begin{pmatrix} 2 \\ -5 \end{pmatrix}$.

2   If points A and B have position vectors **a** and **b**, respectively, then the midpoint M of AB has position vector $\mathbf{m} = \frac{1}{2}(\mathbf{a} + \mathbf{b})$.

Using points A and B from step 1 above, the midpoint of AB has position vector

$\frac{1}{2}(\mathbf{a} + \mathbf{b}) = \frac{1}{2}\begin{pmatrix} 4 \\ 8 \end{pmatrix} + \begin{pmatrix} 6 \\ 3 \end{pmatrix} = \frac{1}{2}\begin{pmatrix} 10 \\ 11 \end{pmatrix} = \begin{pmatrix} 5 \\ 5\frac{1}{2} \end{pmatrix}$. So the midpoint is located at $(5, 5\frac{1}{2})$.

For points C(2,8) and D(7,−4) find $\vec{CD}$ and the midpoint of CD.

D   The magnitude of a vector

1   Any vector not parallel to the co-ordinate axes can be thought of as the hypotenuse of a right-angled triangle. Pythagoras' rule gives its length.

The length of a vector x is called its magnitude or modulus and is written $|x|$.

The magnitude of a vector $\begin{pmatrix} a \\ b \end{pmatrix} = \sqrt{a^2 + b^2}$.

Vector $\vec{MN} = \begin{pmatrix} -2 \\ 8 \end{pmatrix}$.

What is the distance from M to N?

2   For example, the size of the vector $\begin{pmatrix} 3 \\ -4 \end{pmatrix}$ is $\sqrt{3^2 + {-4}^2} = \sqrt{9 + 16} = \sqrt{25} = 5$ units.

PRACTICE

1   In this question, $\mathbf{a} = \begin{pmatrix} 1 \\ 5 \end{pmatrix}$, $\mathbf{b} = \begin{pmatrix} 7 \\ 0 \end{pmatrix}$, $\mathbf{c} = \begin{pmatrix} -2 \\ 2 \end{pmatrix}$, $\mathbf{d} = \begin{pmatrix} 0 \\ -3 \end{pmatrix}$, $n = 3$, $m = -2$.

For each part, (i) simplify the expression if possible, (ii), write the resultant as a column vector, (iii) find its magnitude, correct to 2 d.p. where necessary.

(a) $\mathbf{a} + \mathbf{b}$ (b) $4\mathbf{c}$
(c) $2\mathbf{d} + \mathbf{b} - \mathbf{d}$ (d) $n\mathbf{a} + m\mathbf{a}$
(e) $\mathbf{a} + 3\mathbf{d} - \mathbf{c} - 2\mathbf{a}$

2   On a co-ordinate grid, point J is at (2, 5) and K is at (6, −1). Find as column vectors:

(a) the position vector of J   (b) the position vector of K
(c) $\vec{JK}$ (d) $\vec{KJ}$
(e) the position vector of the
      midpoint of JK

You can think of the midpoint as the 'mean' of the position vectors.

THE BARE BONES

➤ Any kind of motion (such as a journey) can be represented on a graph.

➤ The two main types of motion graph are those in which:
Position ('displacement') is plotted against time;
Velocity is plotted against time.

A Displacement

Q Can you explain the difference between distance and displacement to someone else? Use a car, bus or rail journey in your locality as an example.

KEY FACT

1 A bus travels eastwards from Rhyl to Prestatyn, a distance of 5 miles, then back through Rhyl to Abergele, which is 5 miles west of Rhyl.

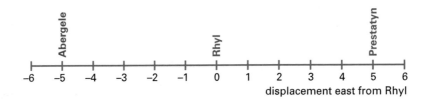

2 If east is the positive direction, then by the time the bus reaches Abergele:
• it has travelled a distance of 15 miles;
• its displacement from Rhyl is −5 miles.

> Distance has only magnitude: there is no direction associated with it. Displacement has both a magnitude and a direction.

B Displacement – time graphs

Remember
When drawing displacement-time graphs, displacement is on the vertical axis and time is on the horizontal axis.

Q Can you explain why the displacement-time graph of a real journey can't have sharp 'corners' like this one?

1 The bus journey from example 2 above might look like this on a graph:

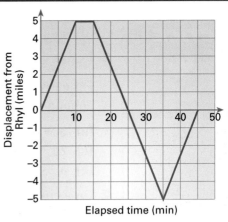

2 You can work out the velocity of the bus from the graph. Velocity is to speed as displacement is to distance: it has magnitude and direction. The velocity is given by the gradient of the graph. On the first section of the journey,

$$\text{gradient} = \frac{\text{increase in displacement}}{\text{increase in time}} = \frac{5 \text{ miles}}{10 \text{ min}} = 0.5 \text{ miles/min} = 30\text{mph}$$

3 On the second leg of the journey, the velocity is −30 miles/hour, as the displacement is decreasing.

> Velocity is a measure of how quickly displacement changes.

4 To estimate the velocity at a point on a curved graph, draw the tangent to the graph at that point, then calculate its gradient.

C Velocity – time graphs

1 This graph shows the journey of a tube train between two stations.

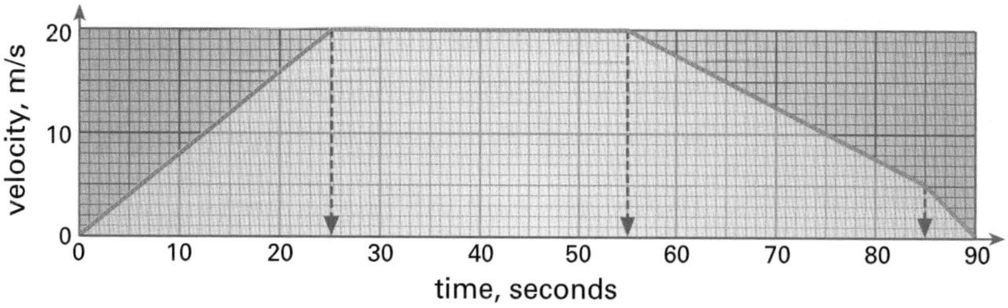

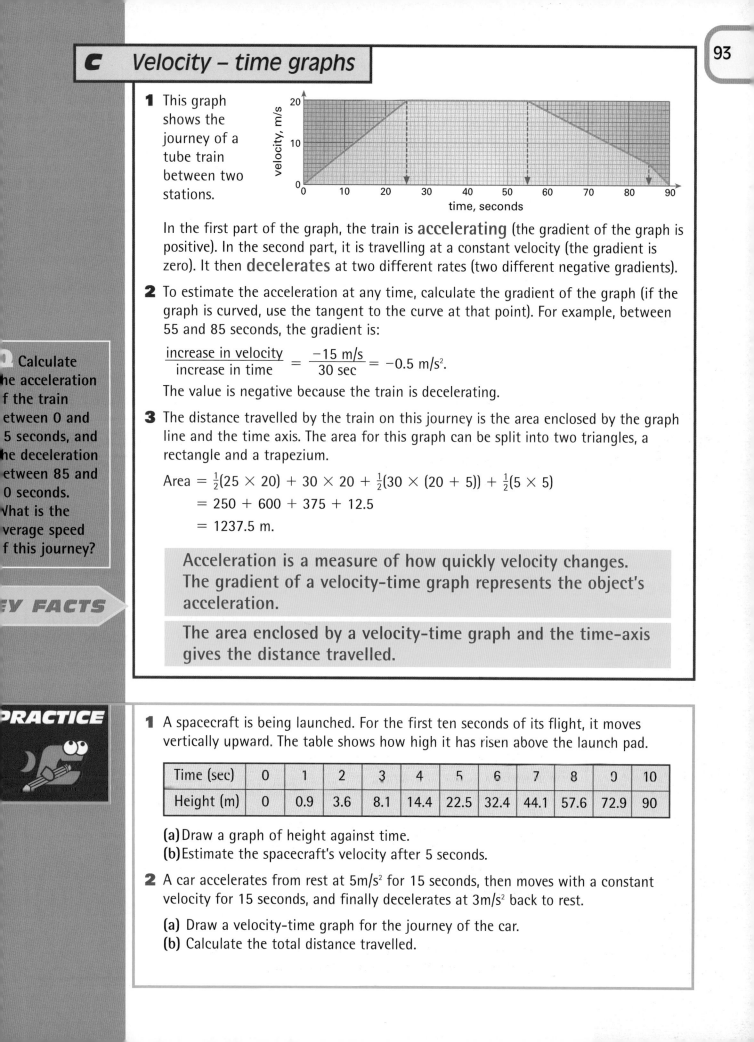

In the first part of the graph, the train is **accelerating** (the gradient of the graph is positive). In the second part, it is travelling at a constant velocity (the gradient is zero). It then **decelerates** at two different rates (two different negative gradients).

2 To estimate the acceleration at any time, calculate the gradient of the graph (if the graph is curved, use the tangent to the curve at that point). For example, between 55 and 85 seconds, the gradient is:

$$\frac{\text{increase in velocity}}{\text{increase in time}} = \frac{-15\text{ m/s}}{30\text{ sec}} = -0.5\text{ m/s}^2.$$

The value is negative because the train is decelerating.

3 The distance travelled by the train on this journey is the area enclosed by the graph line and the time axis. The area for this graph can be split into two triangles, a rectangle and a trapezium.

$$\begin{aligned}\text{Area} &= \tfrac{1}{2}(25 \times 20) + 30 \times 20 + \tfrac{1}{2}(30 \times (20 + 5)) + \tfrac{1}{2}(5 \times 5)\\ &= 250 + 600 + 375 + 12.5\\ &= 1237.5\text{ m.}\end{aligned}$$

Q Calculate
he acceleration
f the train
etween 0 and
5 seconds, and
he deceleration
etween 85 and
0 seconds.
Vhat is the
verage speed
f this journey?

EY FACTS

Acceleration is a measure of how quickly velocity changes. The gradient of a velocity-time graph represents the object's acceleration.

The area enclosed by a velocity-time graph and the time-axis gives the distance travelled.

PRACTICE

1 A spacecraft is being launched. For the first ten seconds of its flight, it moves vertically upward. The table shows how high it has risen above the launch pad.

Time (sec)	0	1	2	3	4	5	6	7	8	9	10
Height (m)	0	0.9	3.6	8.1	14.4	22.5	32.4	44.1	57.6	72.9	90

(a) Draw a graph of height against time.
(b) Estimate the spacecraft's velocity after 5 seconds.

2 A car accelerates from rest at 5m/s^2 for 15 seconds, then moves with a constant velocity for 15 seconds, and finally decelerates at 3m/s^2 back to rest.

(a) Draw a velocity-time graph for the journey of the car.
(b) Calculate the total distance travelled.

Circle geometry

THE BARE BONES

➤ There are a number of angle facts connected with circles that you need to know. It is possible to deduce most of them from other facts, but this wastes time in an exam.

➤ There are two of these facts you need to be able to prove.

A Tangents and chords

Q Can you explain how to construct the perpendicular bisector of a line?

1 The perpendicular bisector of a chord passes through the centre of the circle.

2 The two tangents drawn from a point to a circle are equal.

B Angles in semicircles, at the centre, or in the same segment

1 The angle in a semicircle is a right angle.

2 The angle subtended by any arc at the centre of a circle is twice the angle subtended by the same arc at any point on the remaining part of the circumference. In this diagram, this means that angle x is double the size of angle y.

KEY FACT

> This is a statement you need to be able to prove.

To do this, draw in the radius OB. This divides quadrilateral OABC into two isosceles triangles. The base angles of △OAB have been labelled p, and in △OBC, q. So $y = p + q$.

Remember
Angles that add up to 180° are called supplementary.

Proof
$\angle AOB = 180° - 2p$ (angles in a triangle total 180°)
$\angle BOC = 180° - 2q$ (angles in a triangle total 180°)
$\quad x = 360° - (\angle AOB + \angle BOC)$ (angles at a point total 360°)
$\quad\quad = 360° - (180° - 2p + 180° - 2q)$
$\quad\quad = 360° - (360° - 2p - 2q)$
$\quad\quad = 2p + 2q$
$\quad\quad = 2y.$

This proves the result.

3 Angles in the same segment are equal.

Q Can you explain the meaning of the word 'subtend'?

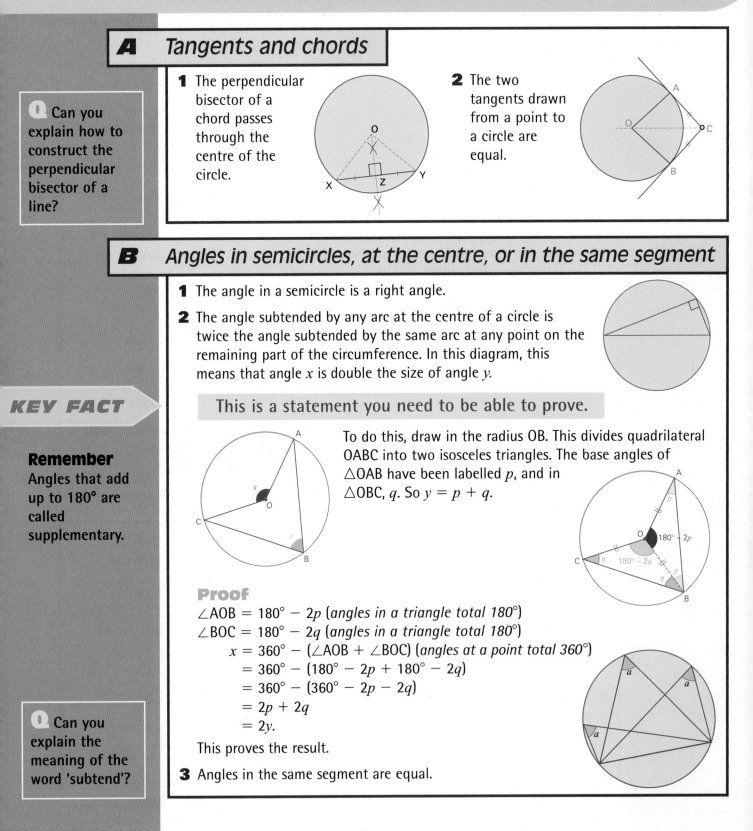

C Cyclic quadrilaterals

1 Opposite angles in a cyclic quadrilateral
are supplementary (they add up to 180°).

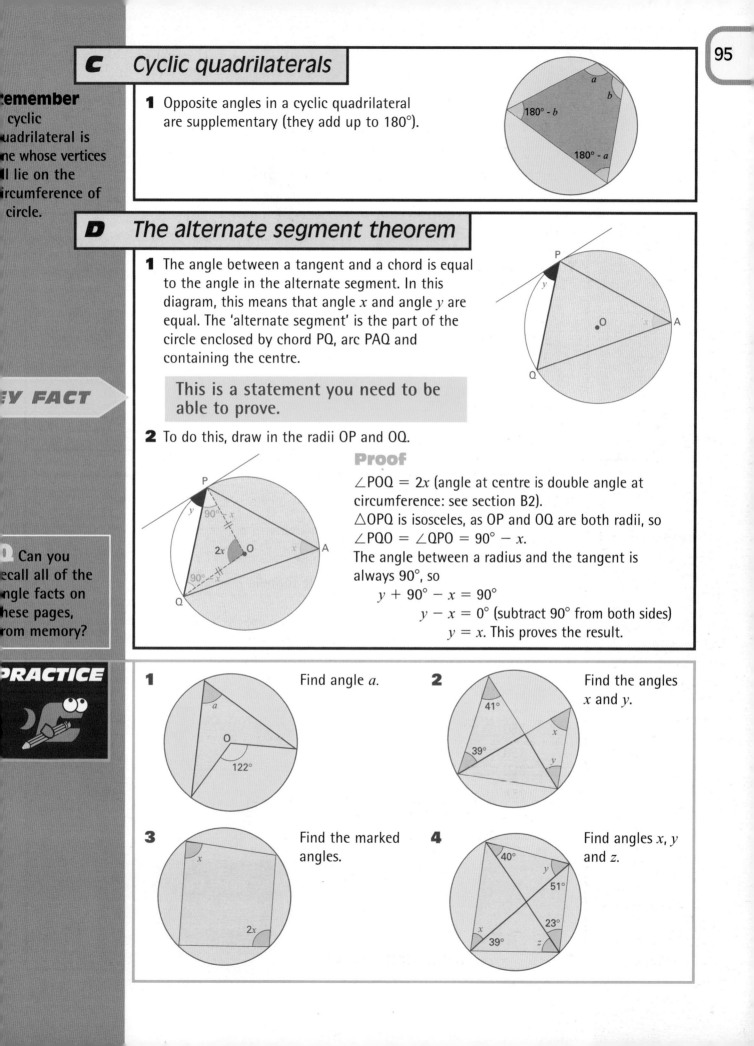

D The alternate segment theorem

1 The angle between a tangent and a chord is equal
to the angle in the alternate segment. In this
diagram, this means that angle x and angle y are
equal. The 'alternate segment' is the part of the
circle enclosed by chord PQ, arc PAQ and
containing the centre.

This is a statement you need to be
able to prove.

2 To do this, draw in the radii OP and OQ.

Proof

$\angle POQ = 2x$ (angle at centre is double angle at
circumference: see section B2).
$\triangle OPQ$ is isosceles, as OP and OQ are both radii, so
$\angle PQO = \angle QPO = 90° - x$.
The angle between a radius and the tangent is
always 90°, so
$$y + 90° - x = 90°$$
$$y - x = 0° \text{ (subtract 90° from both sides)}$$
$$y = x. \text{ This proves the result.}$$

PRACTICE

1 Find angle a.

2 Find the angles
x and y.

3 Find the marked
angles.

4 Find angles x, y
and z.

Circles: arcs and sectors

THE BARE BONES

> The length of an arc is a fraction of the circumference of a circle, e.g. a 90° arc is a quarter of the circumference.
> The area of a sector is a fraction of the area of a circle, e.g. a 60° sector is one-sixth of the area of the whole circle.

A Fractions of a circle

1 A semicircle is half of a circle.
A quadrant is a quarter.

KEY FACT

In general, an arc or sector subtending $x°$ at the centre forms a fraction $\frac{x}{360}$ of the circumference or area.

Q What fraction of a circle is a sector subtending 45° at the centre?

B Length of an arc

Remember
You may be asked to leave the answer 'exact' in terms of π.

1 The whole circumference of a circle of diameter d is $C = \pi d$.

The length L of an arc is a fraction of this.

In the diagram above, this would give the formula $L = \frac{x}{360} \times \pi d$.

If the radius is given, use $L = \frac{x}{360} \times 2\pi r$.

2 Find the arc length between M and N in this diagram.

Here $x = 150°$ and $r = 10$ cm.

$$L = \frac{x}{360} \times 2\pi r$$

$$= \frac{150}{360} \times 2\pi \times 10$$

$$= \frac{25\pi}{3} \text{ (if the answer is to be left in terms of } \pi\text{)}$$

$$= 26.2 \text{ cm to 3 s.f.}$$

Q What is the perimeter of this sector?

Calculate and jot down the circumference or area of the whole circle. This helps you check your answer and could score method marks if the answer is wrong.

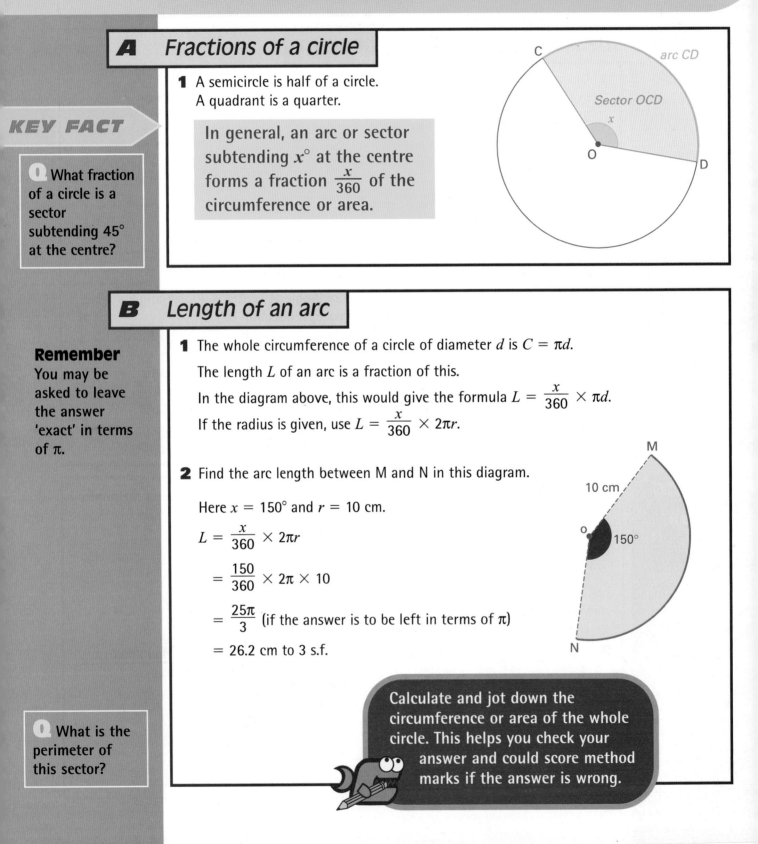

C Area of a sector

1 The area of a circle of radius r is $A = \pi r^2$. The area of a sector is a fraction of this.

> In the diagram in section A, p96, this would give the formula $A = \dfrac{x}{360} \times \pi r^2$

2 Find the area of sector OST in this diagram.

Here $x = 24°$ and $r = 15$ m.

$A = \dfrac{x}{360} \times \pi r^2$

$= \dfrac{24}{360} \times \pi \times 15^2$

$= \dfrac{1}{15} \times \pi \times 225$

$= 15\pi$ (if the answer is to be left in terms of π)

$= 47.1$ m^2 to 3 s.f.

In terms of π, what is the area of a 60° sector from the same circle?

1 Find the lengths of the following arcs. Give your answers to 3 s.f.

(a) 50° 4 cm

(b) 12° 12 m

(c) 41 cm 225°

2 Find the area of each sector. Give your answers to 3 s.f.

(a) 21 cm 40°

(b) 320° 8 mm

(c) 40 km 125°

3 A decorative tile is in the shape of a quarter-circle with radius 11 cm. What is the perimeter of the tile, to the nearest millimetre?

4 What is the area of this shape? The outer circle has a diameter of 1 m. Give your answer in square metres, to 2 d.p.

135° 25 cm

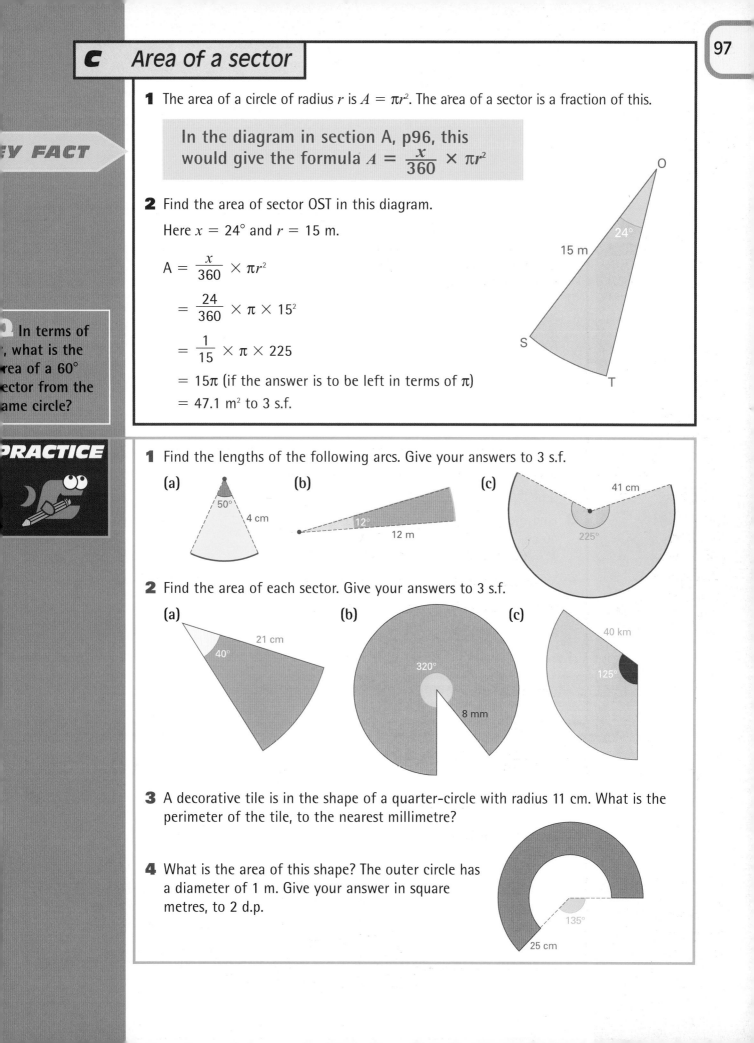

THE BARE BONES

➤ Some of the formulae used in this section are printed at the front of your examination paper. However, it's much safer not to rely on these, but to learn as many as you can.

A Prisms and cylinders

Area A

Perimeter P

h

1 If a prism has height h and a base with area A and perimeter P, then the following formulae are true:

KEY FACT

> Volume of prism $V = Ah$
> Surface area $S = 2A + Ph$.

2 A cylinder is a prism with a circular base. If the circle has radius r, then the base has area πr^2 and perimeter (circumference) $2\pi r$.

The prism formulae thus become:

KEY FACT

> $V = Ah = \pi r^2 h$
> $S = 2A + Ph = 2\pi r^2 + 2\pi rh$, which can be factorised to $2\pi r(r + h)$.

Remember
A prism is a solid with a uniform cross-section.

3 Find the volume and surface area of a cylinder with base radius 10 cm and height 40 cm.

$V = \pi r^2 h$
$\quad = \pi \times 10^2 \times 40$
$\quad = 4000\pi$ (if the answer is to be left in terms of π)
$\quad = 12\ 566\ \text{cm}^3$ (to nearest whole cm³).
$S = 2\pi r^2 + 2\pi rh$
$\quad = 2 \times \pi \times 10^2 + 2 \times \pi \times 10 \times 40$
$\quad = 200\pi + 800\pi$
$\quad = 1000\pi$ (if the answer is to be left in terms of π)
$\quad = 3141.6\ \text{cm}^2$ (to 1 d.p.)

Q How does a cuboid fit this pattern?

B Pyramids and cones

Apex

1 If a pyramid has height h and a base with area A, then:

KEY FACT

> Volume of pyramid $V = \frac{1}{3}Ah = \dfrac{Ah}{3}$

There is no general formula for the surface area.

Note that the perpendicular height h is 'dropped' from the apex of the pyramid at right angles to the base. As with heights of triangles, it's possible that this might fall outside the base itself, but this doesn't matter.

h

Area A

B

2 A cone is a pyramid with a circular base. If the circle has radius r, then the base has area πr^2. The pyramid volume formula becomes:

$$V = \tfrac{1}{3}\ Ah = \tfrac{1}{3}\pi r^2 h.$$

Find the volume and surface area of a cone with base radius 12 cm and height 6 cm.

3 There is a formula for the surface area of a cone. This involves the slant height, usually denoted by l.

$S = \text{base} + \text{curved surface} = \pi r^2 + \pi r l$, which factorises to $\pi r(r + l)$.

Note that the slant height is part of a right-angled triangle, and so by Pythagoras' rule, $l = \sqrt{r^2 + h^2}$.

C Spheres

1 If the radius of a sphere is r:

> The volume is $V = \tfrac{4}{3}\pi r^3$
>
> The surface area is $S = 4\pi r^2$

KEY FACT

What is the surface area of this sphere?

2 The volume of a sphere is 1200 cm³: calculate its radius.

$$V = \tfrac{4}{3}\pi r^3 = 1200$$

So $\pi r^3 = \tfrac{3}{4} \times 1200 = 900$

So $r^3 = \dfrac{900}{\pi}$

$r = \sqrt[3]{\dfrac{900}{\pi}} = 6.59$ cm to 2 d.p.

Always quote the formula in algebra and then show that you can substitute into the formula.

PRACTICE

1 A pyramid has a square base, 10 cm on each side, and a perpendicular height of 15 cm. Find its volume.

2 Find the surface area of a cylinder of base radius 14 cm and height 30 cm.

3 Find the volume of a sphere of diameter 30 cm.

4 A cone of base radius 9 cm has a slant height of 15 cm. Find the volume of the cone.

5 A sphere has volume 288π mm³. Calculate its surface area.

6 Calculate the total surface area of a triangular prism with the following description:

• its base is a right-angled triangle with base 12 cm and height 5 cm;

• its perpendicular height is 20 cm.

Remember
You can use the value π on your calculator.

Cumulative frequency

THE BARE BONES

➤ Think of cumulative frequency as a running total. The frequencies are added as you go along.

➤ The median, quartiles and interquartile range can be estimated from the curve.

A Drawing a cumulative frequency diagram

Remember
The interquartile range is where the middle 50% of the data is located.

A cumulative frequency curve often has an S-shape.

1 It is split into a number of parts:

- The **median** is at the 50% point and is known as Q_2.
- The **upper quartile** is at the 75% point and is known as Q_3 or *UQ*.
- The **lower quartile** is at the 25% point and is known as Q_1 or *LQ*.
- The upper and lower quartiles are used to find the central 50% of the distribution. This is known as the **interquartile range**.

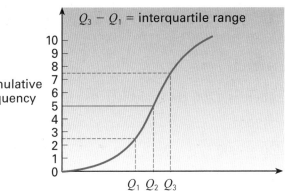

Age 0–9 means '0–9 years 364 days', that is, nearly 10. Age is a difficult variable — be careful in the exam.

2 This table illustrates the **age distribution** in a village of 720 people:

age	0–9	10–19	20–29	30–39	40–49	50–59	60–69	70–79	80–100
frequency	48	72	65	120	153	50	72	96	44

You can draw a cumulative frequency diagram to illustrate this data:

age (less than)	10	20	30	40	50	60	70	80	100
cumulative frequency	48	120	185	305	458	508	580	676	720

The cumulative frequency table is built up by adding the frequencies to what came before, so the 120 in the second cell is 48 + 72 (from the first table).

The 185 in the third cell is the 120 from the second cell + 65 from the first table, and so on.

There is a sketch of how this diagram should look on the opposite page.

Q Can you describe the interquartile range?

B   *Using the diagram*

1 This is a sketch of the curve from the information in the table opposite.
It has age plotted against **cumulative frequency**.

Q_1, Q_2 and Q_3 go across to the curve and then down to the age-axis. So the value of Q_1 is the value the downward line makes with the age-axis, and not 25%.

2 You could draw your own accurate graph using the information in the tables. You can use this sketch or your graph to find the **median** and the **interquartile range**.

- **Median:**

The median is Q_2, so draw a line from the 50% mark on the cumulative frequency axis, that is half-way through the distribution, across to the curve and then down to the age-axis.

This value is the median.

- **Interquartile range:**

Draw a line from Q_3 and Q_1 across to the curve and then down to the age-axis.

The interquartile range $= Q_3 - Q_1$.

Q_3 on this sketch is 64, Q_1 is 30, so the interquartile range is $64 - 30 = 34$.

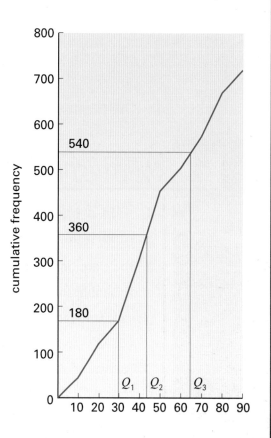

Why are the
uartiles not 25,
0 and 75
ears?

This is the age distribution of the population of a country:

age	number of people (Millions)
under 10	16
10-19	12
20-29	17
30-39	16
40-49	15
50-69	10
70-89	4

1 Use the information in the table to draw a cumulative frequency table.

2 Use the information in the table to draw the cumulative frequency curve.

3 Estimate the median and interquartile range.

Histograms

➤ When data is in equal class intervals, the heights of the bars are proportional to frequency, and a histogram resembles an ordinary bar chart.
➤ The frequency density of a histogram is calculated as $\dfrac{\text{frequency}}{\text{class width}}$.

A Constructing histograms

> **KEY FACT**

In a histogram, the areas of the rectangles are proportional to the frequencies they represent.

1 There are times when data that is collected forms class intervals with unequal widths.

This table shows the results of a survey of workers' journey times to work:

Duration of Journey (t minutes)	$0 \leqslant t < 10$	$10 \leqslant t < 15$	$15 \leqslant t < 20$	$20 \leqslant t < 25$	$25 \leqslant t < 30$
Frequency	12	14	20	10	4

Remember
The class width is the difference between the upper and lower class boundaries.

2 Now calculate the **frequency density** of each class interval. This is done by using the equation:

$$\text{frequency density} = \frac{\text{frequency}}{\text{class width}}$$

In this case, the frequency density represents the number of workers *per minute of journey time*.

This means the table now reads:

Duration of Journey (t minutes)	$0 \leqslant t < 10$	$10 \leqslant t < 15$	$15 \leqslant t < 20$	$20 \leqslant t < 25$	$25 \leqslant t < 30$
Frequency density	$12 \div 10 = 1.2$	$14 \div 5 = 2.8$	$20 \div 5 = 4$	$10 \div 5 = 2$	$4 \div 5 = 0.8$

3 These figures give the heights of the bars in the histogram:

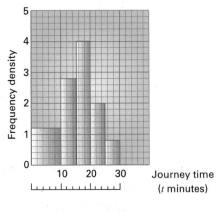

Q Can you explain the meaning of frequency density to another person?

B Interrogating histograms

1 As with all visual ways of displaying data, much useful information can be extracted from a histogram.

> To find the actual frequencies, multiply the frequency density by the class width.

This histogram shows the lifespan of worms in a wormery, before they were sold on:

Lifetime (L weeks)	Frequency density	Frequency
$0 \leqslant L < 4$	16	
$4 \leqslant L < 6$	25	
$6 \leqslant L < 8$	20	
$8 \leqslant L < 10$	10	
$10 \leqslant L < 12$	25	
$12 \leqslant L < 16$	5	

The frequency density can be read straight from the graph.

For example, for the class $0 \leqslant L < 4$, frequency = freq density × class width = 16 × 4 = 64 worms.

2 Calculate the rest of the frequencies for the table above. You should find that the total frequency is 244 worms.

Can you explain how to find the frequencies when the frequency density and class widths are known?

PRACTICE

1 In a survey at Valley College, students were asked how long it had taken them to travel in that morning. This table shows the results of the survey. There are 50 students at the college.

Duration of journey (t minutes)	Frequency
$0 \leq t < 5$	113
$5 \leq t < 10$	114
$10 \leq t < 15$	220
$15 \leq t < 20$	120
$20 \leq t < 25$	110
$25 \leq t < 30$	141

(a) Draw a histogram illustrating this data.
(b) How many students were absent on the day of the survey?

2 Graham grows tomatoes. This histogram shows the heights of the plants in his greenhouses.

(a) Draw a table to record the classes and frequency density.
(b) Add another column to the table for the frequencies, then calculate them to complete the column.
(c) How many tomato plants does Graham have?

Dispersion

THE BARE BONES

➤ The dispersion of a set of data is a measurement of how 'spread out' the data is.

➤ Different measures of dispersion are the range, interquartile range, variance and standard deviation.

A Variance and standard deviation

Remember
The range is the difference between the highest and lowest data values. The interquartile range is where the middle 50% of the data lies.

1 Ursula grows roses. In a test she used different composts and achieved these results for growth per week, measured in mm. She wanted to know which compost was better.

	Week 1	Week 2	Week 3	Week 4	Week 5
Compost type 1	3	12	8	19	8
Compost type 2	2	10	10	10	18

2 For both compost types, the mean growth was 10 mm, and the range of values was 16 mm.

It looks as though type 2 was more consistent, as three of its results equalled the mean.

Remember
The mean of quantity x is written $\bar{x}$ (pronounced x-bar).

3 To compare the composts, look at the deviation of each result from the mean score.

$x - \bar{x}$	Week 1	Week 2	Week 3	Week 4	Week 5
Compost type 1	−7	2	−2	9	−2
Compost type 2	−8	0	0	0	8

The mean deviation for each type is 0. The positive and negative values cancel each other out. This is true for any data set.

4 Squaring the deviations makes all the results positive.

$(x - \bar{x})^2$	Week 1	Week 2	Week 3	Week 4	Week 5
Compost type 1	49	4	4	81	4
Compost type 2	64	0	0	0	64

Q Can you explain how to find the standard deviation of a set of data to someone else?

The mean of these values is known as the **variance** of the data.
It is $142 \div 5 = 28.4$ for type 1 and $128 \div 5 = 25.6$ for type 2.

5 The square root of the variance is called the **standard deviation** of a data set, and is an important measure of dispersion. For type 1 it is 5.3 mm and for type 2, 5.1 mm. So the results for type 1 have a slightly greater spread than type 2.

A

To calculate the standard deviation of a set of data (x):
- Calculate the mean ($\bar{x}$).
- Calculate the deviation of each data item from the mean:
  ($x - \bar{x}$).
- Square the deviations ($x - \bar{x}$)^2.
- Calculate the mean of the squared deviations.
- Find the square root of this mean.

B *In the exam*

1 There are formulae given for standard deviation at the front of the exam paper. They will normally look something like this:

> Standard deviation for a set of numbers
> $x_1, x_2, \dots x_n$, having a mean of $\bar{x}$ is given by
>
> $$s = \sqrt{\frac{\Sigma(x - \bar{x})^2}{n}} \quad \text{or} \quad s = \sqrt{\frac{\Sigma x^2}{n} - \left\{\frac{\Sigma x}{n}\right\}^2}$$

They use the Greek sigma (Σ) to stand for 'sum'. It just means 'find the total of...'.

2 Most scientific calculators have facilities to enter data and calculate statistics such as the mean, variance and standard deviation. You may well be asked to use these functions, and though you may be able to calculate the statistics manually, being familiar with your calculator's special functions could save a lot of time.

Describe the expression inside the square root, in the second formula in the box.

PRACTICE

1 Find the standard deviation of each of the following sets of data:

(a) 1 4 8 8

(b) 2 5 5 6 7

(c) 1 2 5 6 7 7

2 Calculate the mean and the standard deviation of the following sets of values:

(a) 33 36 38 40 45

(b) 12 14 16 18 22 26

3 This set of marks was obtained by 15 students in an English GCSE examination:

77 78 55 89 88 55 89 97 64 68 67 58 59 78 76

(a) Find the mean of this set of marks.

(b) Calculate the standard deviation for this set of marks.

(c) All of the students had their marks raised by five marks, for good spelling. For the new set of marks, find the mean and the standard deviation.

Probability: the OR rule

THE BARE BONES

➤ If two outcomes are mutually exclusive, it means they cannot happen at the same time.

➤ The probability of something happening is 1 minus the probability of it not happening.

A Probability

KEY FACT

1 Mutually exclusive outcomes cannot happen at the same time.

- If one outcome occurs, it prevents the other outcome from happening.

- The spin of a coin will give either a head or a tail. Both head and tail cannot appear at the same time, so they are mutually exclusive.

KEY FACT

When the probabilities of mutually exclusive outcomes are added together, the answer is 1.

The probability that something <u>will not happen</u> is: 1 minus the probability of it <u>happening</u>.

2 The probability that it will NOT rain tomorrow in Caerleon is $\frac{3}{10}$. What is the probability that it WILL rain tomorrow?

The probability that it will rain in Caerleon tomorrow is 1 minus the probability that it will not rain, that is:

$1 - \frac{3}{10} = \frac{7}{10}$

With a probability this high, it seems likely to rain, so it would be a good idea to take an umbrella in this area!

Q Use this method to answer this question. A playing card is selected at random from a pack of 52. What is the probability of not selecting a King?

Make sure you know how to calculate with fractions.

B   The 'OR' rule

When two outcomes, X and Y, of an event are mutually exclusive, the probability of one or the other occurring is the sum of their probabilities: P(X OR Y) = P(X) + P(Y).

Remember
You can only use the 'OR' rule if the outcomes are mutually exclusive.

Q Is the probability of selecting a King or a Jack greater than the probability of selecting a King or a Queen in a standard pack of playing cards?

1 A disc is selected at random from a bag containing 5 blue discs, 6 yellow discs and 3 white discs. What is the probability of selecting either a blue disc or a white disc?

The two outcomes are **mutually exclusive.**

P (a blue disc OR a white disc)$= $ P (blue) + P (white)

$$= \tfrac{5}{14} + \tfrac{3}{14}$$

$$= \tfrac{8}{14} = \tfrac{4}{7}$$

2 A playing card is selected at random from a pack.

What is the probability that it is a Jack or a King?

$$P \text{ (selecting a Jack)} = \tfrac{4}{52}$$

$$P \text{ (selecting a King)} = \tfrac{4}{52}$$

$$P \text{ (selecting a Jack OR a King)} = \tfrac{4}{52} + \tfrac{4}{52} = \tfrac{8}{52}$$

$$= \tfrac{2}{13}$$

PRACTICE

1 The probability of a driver passing her driving test is 0.85. What is the probability of her failing her driving test?

2 When a fair die is rolled, what is the probability of NOT getting a:

(a) 6?   (b) prime number?

3 A bag contains lots of coloured discs. They are white, brown, black and pink. The probabilities of picking each colour of disc are shown in the table:

colour	white	brown	black	pink
probability	0.4	0.3	0.17	

(a) What is the probability of picking a pink disc?

(b) What is the probability of picking a disc that is NOT brown?

4 The probability that I will drive to work tomorrow is 0.95, the probability that I will walk is 0.05.

(a) What is the probability that I will take the bus?

(b) What is the probability that I will drive or walk to work?

5 A bag contains 7 red counters, 8 blue counters and 11 orange counters. One is picked at random.

(a) What is the probability that it is red or blue?

(b) What is the probability that is it not blue?

Probability: the AND rule

➤ Independent events occur when the outcome of one event does not depend on another event's outcome occurring.

➤ Tree diagrams are a useful way of demonstrating all the possible outcomes from sets of independent events, and their probability.

A   The 'AND' rule

Q If you flip a coin, then flip it again, are the results independent?

1 If the occurrence of one event is unaffected by the occurrence of another event, then the events are said to be **independent**. When two events are independent, the probability of one event and the other event occurring is the product of their probabilities.

KEY FACT

> This is the 'AND' rule:
> $P(X \text{ AND } Y) = P(X) \times P(Y)$

2 Two coins are spun through the air at the same time. Find the probability of obtaining a head and a head.

$$P(\text{Head AND Head}) = P(H) \times P(H)$$
$$= \tfrac{1}{2} \times \tfrac{1}{2}$$
$$= \tfrac{1}{4}$$

3 A coin is spun through the air and a fair dice is rolled. What is the probability of:

(a) obtaining a head on the coin? $= \tfrac{1}{2}$

(b) obtaining a 6 on the dice? $= \tfrac{1}{6}$

(c) obtaining a head on the coin AND a 6 on the dice? $= \tfrac{1}{2} \times \tfrac{1}{6} = \tfrac{1}{12}$

B   Tree diagrams with two branches

KEY FACT

> To read a tree diagram, you need to start at the stem and work along the branches of the tree.

Remember
All the fractions at the end of the pairs of branches should add together to give an answer of 1.

1 A box contains 5 blue discs and 3 red discs. A disc is selected at random and replaced. A second disc is then selected at random. What is the probability that both discs are blue?

To find the probability of a blue followed by a blue, read along the top branch. The probability of a blue and a blue $= \tfrac{5}{8} \times \tfrac{5}{8} = \tfrac{25}{64}$.

P(blue) $\tfrac{5}{8}$ P(blue) $\tfrac{5}{8}$ P(red) $\tfrac{3}{8}$

P(red) $\tfrac{3}{8}$ P(blue) $\tfrac{5}{8}$ P(red) $\tfrac{3}{8}$

2 A box contains 5 blue discs and 3 red discs. A disc is selected at random and NOT replaced. A second disc is then selected at random.

The probability that both discs are blue $= \tfrac{5}{8} \times \tfrac{4}{7} = \tfrac{20}{56} = \tfrac{5}{14}$.

The probability that one is blue and one is red is the probability of a red and a blue, **or** a blue and a red.

This is an example that uses the AND rule as well as the OR rule. P(a red and a blue) or P(a blue and a red) $= (\tfrac{3}{8} \times \tfrac{5}{7}) + (\tfrac{5}{8} \times \tfrac{3}{7}) = \tfrac{15}{28}$.

P(blue) $\tfrac{5}{8}$ P(blue) $\tfrac{4}{7}$ P(red) $\tfrac{3}{7}$

P(red) $\tfrac{3}{8}$ P(blue) $\tfrac{5}{7}$ P(red) $\tfrac{2}{7}$

Q What is the probability that you will draw two discs of the same colour?

c Tree diagrams with more than 2 branches

1 A bag containing 10 blue balls and 7 white balls is placed on a table.

2 Draw a tree diagram to show all possible combinations for two selections where the balls are replaced after each selection.

P(blue) $\frac{10}{17}$
P(white) $\frac{7}{17}$
P(blue) $\frac{10}{17}$
P(white) $\frac{7}{17}$
P(blue) $\frac{10}{17}$
P(white) $\frac{7}{17}$

3 What would the tree diagram look like if the process were repeated with 5 extra red balls added to the bag, and if the ball were not replaced after the first selection?

P(blue) $\frac{10}{22}$
P(white) $\frac{7}{22}$
P(red) $\frac{5}{22}$
P(blue) $\frac{9}{21}$
P(white) $\frac{7}{21}$
P(red) $\frac{5}{21}$
P(blue) $\frac{10}{21}$
P(white) $\frac{6}{21}$
P(red) $\frac{5}{21}$
P(blue) $\frac{10}{21}$
P(white) $\frac{7}{21}$
P(red) $\frac{4}{21}$

Q What is the probability that you will draw two balls of the same colour?

1 A card is drawn from a fair pack and a coin is spun through the air. What is the probability that:

(a) it is a black card?

(b) the coin shows tails?

(c) the card is black AND the coin shows tails?

2 A playing card is picked at random from a pack and then replaced. The pack is shuffled before a second card is taken. What is the probability that:

(a) both cards are Clubs?

(b) both cards are Kings?

(c) both cards are not picture cards?

3 A bag contains 5 red counters and 4 blue counters. A counter is drawn at random and replaced before another counter is drawn again. Draw a tree diagram to show all the possible outcomes. What is the probability that:

(a) two red counters are drawn?

(b) a red and a blue counter are drawn?

NUMBER

There are many specific number questions on both papers – however, many of the facts and skills you learn are used in other branches of mathematics. The questions in this section test 'pure' number skills.

Specimen question 1

a) i) Show that $\sqrt{5} \times \sqrt{15} = 5\sqrt{3}$. (1 mark)

ii) Expand and simplify $(\sqrt{5} + \sqrt{15})^2$. (3 marks)

b) Is ABCD a rectangle?
Show your working clearly. (4 marks)

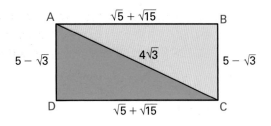

Model answer 1

a) i) $\sqrt{5} \times \sqrt{15}$

$= \sqrt{5} \times \sqrt{5} \times \sqrt{3}$

> Use $15 = 5 \times 3$ within the square root

$= 5\sqrt{3}$

> Combine the two $\sqrt{5}$ to make 5: 1 mark

- You could also have done the question this way:
$\sqrt{5 \times 15} = \sqrt{75} = \sqrt{25 \times 3} = 5\sqrt{3}$.

a) ii) $(\sqrt{5} + \sqrt{15})^2$

$= \sqrt{5}^2 + \sqrt{15}^2 + 2\sqrt{5}\sqrt{15}$

> Use expansion of $(a + b)^2$: 2 marks

$= 5 + 15 + 2 \times (5\sqrt{3})$

$= 20 + 10\sqrt{3}$ | Simplify correctly: 1 mark

- This part uses the algebraic identity $(a + b)^2 = a^2 + 2ab + b^2$.

b) For it to be a rectangle, angle ABC must be a right angle, so by Pythagoras' rule, $AB^2 + BC^2 = AC^2$.

$AC^2 = (4\sqrt{3})^2 = 16 \times 3 = 48$.

> Expand AC^2 correctly: 1 mark

$AB^2 + BC^2 = (\sqrt{5} + \sqrt{15})^2 + (5 - \sqrt{3})^2$

$\qquad = (20 + 10\sqrt{3}) + (25 + 3 - 2 \times 5\sqrt{3})$

> Use the results from (a): 1 mark

$\qquad = (20 + 10\sqrt{3}) + (28 - 10\sqrt{3})$

$\qquad = 48$.

> Simplify $AB^2 + BC^2$ correctly: 1 mark

So $AB^2 + BC^2 = AC^2$, and ABCD is a rectangle.

- The key to this part is realising that, although it's a number question about surds, you need a geometrical technique (Pythagoras' rule) to get you started.

Specimen question 2

On a TV quiz show, teams buzz to answer a 'starter' question worth 10 points.

- If they **interrupt** the quizmaster and give a **wrong** answer, they lose 5 points.
- If they get the starter **right**, they also answer **extra** questions worth 5 points each.
- If they get a starter question **wrong**, but **don't** interrupt, they score no points.

These were the details of one programme.

	New Bridge College	University of Lakeland
Starters correct	16	18
Starters Wrong	13	9
Starters Wrongly Interrupted	5	11
Extra questions	25	26

Who won the game? (3 marks)

Model answer 2

New Bridge College:

$(16 \times 10) + (13 \times 0) - (5 \times 5) + (25 \times 5) =$

$160 + 0 - 25 + 125 = 260$ points | 1 mark |

University of Lakeland:

$(18 \times 10) + (9 \times 0) - (11 \times 5) + (26 \times 5) =$

$180 + 0 - 55 + 130 = 255$ points | 1 mark |

New Bridge College won.

| Answer the question: 1 mark |

- To do this question, calculate the total score for each team.

- This is a Paper 1 question, so the calculations need to be done mentally, or with extra working on the paper. The numbers have been kept easy to help you!

- An alternative (but unorthodox) method is to work out the difference between the teams for each column in the table. This would give Lakeland the following number of points more than New Bridge: $(2 \times 10) + (-4 \times 0) - (6 \times 5) + (1 \times 5) = 20 + 0 - 30 + 5 = -5$ (so Lakeland lost).

Further questions

1 a) Use your calculator to find $\sqrt{7 - 1.5^3}$.

Write down all the figures on your calculator.

b) Write your answer to 4 significant figures.

2 Which of the following fractions is nearest to $\frac{3}{4}$?

$\frac{11}{15}$ $\frac{23}{30}$ $\frac{34}{45}$ $\frac{43}{60}$

3 (a) An average human hair is 0.00005 mm in diameter.

Write this number in standard index form.

(b) One of the hairs from Michelle's head is 18 cm long. Assuming it is of average diameter, how many times longer is it than it is wide? Give your answer in standard index form.

4 What fraction is equivalent to the recurring decimal 0.135353535...?

Answers

1 (a) 1.903943276...

(b) 1.904

2 $\frac{3}{4} = \frac{135}{180} = 0.75$

$\frac{11}{15} = \frac{132}{180} = 0.7333...$

$\frac{23}{30} = \frac{138}{180} = 0.7666...$

$\frac{34}{45} = \frac{136}{180} = 0.7555...$

$\frac{43}{60} = \frac{129}{180} = 0.7166...$

So $\frac{34}{45}$ is closest.

3 (a) 5×10^{-5} mm

(b) $480 \div (5 \times 10^{-5}) = 9.6 \times 10^{6}$ times

4 $\frac{67}{495}$

SHAPE, SPACE AND MEASURES

This part of mathematics covers many different topics, including all work on angles, areas and volumes of shapes, and anything to do with measurement, including speed.

Specimen question 1

ABC is a triangle with a point O located inside it. $\overrightarrow{OA} = \mathbf{a}$, $\overrightarrow{OB} = \mathbf{b}$, $\overrightarrow{OC} = \mathbf{c}$.

The midpoints of AB, BC and CA are J, K, and L, respectively.

(a) Find, in terms of $\mathbf{a}$, $\mathbf{b}$ and $\mathbf{c}$, the vectors representing:

i) $\overrightarrow{AC}$ (1 mark)

ii) $\overrightarrow{JK}$ (3 marks)

(b) Write down, in terms of $\mathbf{a}$, $\mathbf{b}$ and $\mathbf{c}$, the vectors representing $\overrightarrow{KL}$ and $\overrightarrow{LJ}$ (2 marks)

(c) i) Describe the relationship between triangle ABC and triangle JKL. Give reasons for your answer. (2 marks)

ii) The area of triangle JKL is 8.5 cm². What is the area of triangle ABC?   (1 mark)

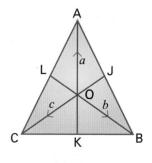

Model answer 1

(a)i) $\vec{AC} = c - a \; (-\vec{OA} + \vec{OC})$: $\boxed{1 \text{ mark}}$

> • This is a fact you should memorise.

ii) $\vec{JK} \quad = (-\vec{OJ} + \vec{OK})$

$\quad = -\frac{1}{2}(a + b) + \frac{1}{2}(b + c)$

> Correct procedure for midpoints: 1 mark

$\quad = -\frac{1}{2}a - \frac{1}{2}b + \frac{1}{2}b + \frac{1}{2}c$

$\quad = -\frac{1}{2}a + \frac{1}{2}c$

$\quad = \frac{1}{2}(c - a)$

> Simplify correctly: 2 marks

> • The location of a midpoint is also a fact you should memorise.

(b) $\vec{KL} = \frac{1}{2}(a - b)$

> Realise that this is like the last part: 1 mark

> • You can do this the 'long' way, but it's much easier just to realise that the procedure is the same as that in (a(ii)), using two other points.

$\vec{LJ} = \frac{1}{2}(b - c)$ $\boxed{1 \text{ mark}}$

(c) i) The vectors for the sides of JKL are half those of the corresponding sides in ABC. JKL is therefore <u>similar to ABC and a quarter of its area.</u>

> 1 mark for each fact

> • You need to spot that $\vec{CB}$ and $\vec{BA}$ are related to the vectors you just found.

ii) $4 \times 8.5 \text{ cm}^2 = \underline{34 \text{ cm}^2}$

> JKL is double the size of ABC, so the area is 4 times larger: 1 mark

> • An enlargement with scale factor s increases the area of a shape by a factor of s^2.

Specimen question 2

The diagram shows a box in the shape of a cuboid. A metal rod 1 m long has been placed in the box with one end at E and the other resting against edge CG, at P.

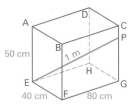

(a) How far below C is the top end of the rod? Give your answer to the nearest centimetre.

(1 mark)

(b) Calculate angle PEG, the angle of inclination of the rod to the horizontal. Give your answer to the nearest degree.

(3 marks)

Model answer 2

(a) From triangle EFG, by Pythagoras' rule,
$EG^2 = d^2 = 40^2 + 80^2 = 8000$

> You either write this down, draw a diagram like the one below, or mark these lengths on the original diagram: 1 mark; then calculate d^2: 1 mark

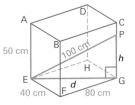

Using Pythagoras' rule in triangle PEG,
$h^2 = EP^2 - d^2 = 100^2 - 8000 = 2000$

> State Pythagoras' rule and substitute values: 1 mark

$h = \sqrt{2000} = 44.721... \text{ cm}$

Required length PC $= 50 - h = 5.278... \text{ cm}$
$= \underline{5 \text{ cm, to nearest cm}}$

> Identify that you need Pythagoras: 1 mark

> • It's important to identify the two triangles you're going to use.
> • In the first triangle, you're going to re-use d^2, so there's no need to calculate d itself.
> • Remember that you're not finding the hypotenuse in triangle PEG, so you subtract values in Pythagoras' rule.

(b) In triangle PEG, tan PEG = opposite ÷ adjacent

State the correct trig ratio and ...

$= h \div d$

$= \sqrt{2000} \div \sqrt{8000}$

Substitute correctly: 1 mark

$= 0.5$

So angle PEG = $\tan^{-1}(0.5)$

Use the inverse trig ratio to calculate an angle

$= 26.565...°$ Calculate the angle

<u>The rod makes an angle of 27° with the horizontal, to the nearest degree.</u>

Give the rounded answer: 1 mark

- The crucial thing here, of course, is to use the right trig ratio!
- Remember to keep the values of trig ratios in your calculator memory ready for the inverse calculation, so there are no rounding errors.

Further questions

1 A circle of radius 8 cm, centre O, has a regular octagon ABCDEFGH positioned so that its vertices are points on the circumference.

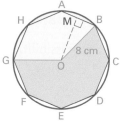

(a) Calculate the area of the minor sector OGHAB. Give your answer in terms of π.

(b) Calculate the perimeter of the minor sector OGHAB. Give your answer in terms of π.

(c) M is the midpoint of AB. By considering triangle OBM, calculate the area of the octagon. Give your answer to the nearest square centimetre.

2 The diagram shows the dimensions of a pentagonal prism.

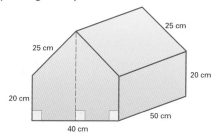

(a) Sketch the net of the prism. Mark on relevant measurements.

(b) Calculate the surface area of the prism.

(c) Calculate the volume of the prism. Give your answer in cubic metres.

Answers

1 **(a)** 24π cm²

(b) $16 + 6\pi$ cm

(c) 181 cm²

2 **(a)**

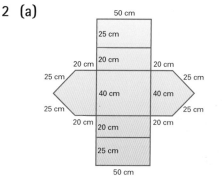

(b) One side = $(40 \times 20) + \frac{1}{2}(40 \times 15) = 1100$ cm²

Rectangles = $130 \times 50 = 6500$ cm²

Total = 8700 cm²

(c) $1100 \times 50 = 55\,000$ cm³ = 0.055 m³

ALGEBRA

Although algebra includes work on co-ordinates and graphs, the main skill you require is to be able to manipulate and transform algebraic expressions. With this skill, you can solve equations, transform formulae and analyse sequences.

Specimen question 1

y is inversely proportional to x^2.

(a) When $x = 5$, $y = 30$.

Find the value of y when $x = 10$.   (3 marks)

Also, x is proportional to the cube root of u

(b) When $y = 120$, $u = 0.1$.

Find the value of u when $x = 5$.   (4 marks)

Model answer 1

(a) $y \propto \dfrac{1}{x^2}$, so $y = \dfrac{k}{x^2}$

Write down the equation of proportionality: 1 mark

So $30 = \dfrac{k}{5^2} = \dfrac{k}{25}$ — Substitute values

$$k = 750$$

Calculate the constant of proportionality: 1 mark

When $x = 10$, $y = \dfrac{750}{x^2}$

$$= \dfrac{750}{100^2}$$ — Substitute for x

$$= \underline{7.5}.$$ — Calculate y: 1 mark

- It's important to follow the initial steps of this answer through very carefully. Don't miss out any steps. Check that your equation matches the statement at the beginning of the question.

(b) If $y = 120$, $120 = \dfrac{750}{x^2}$

Equation of proportionality

$$120x^2 = 750$$ — Rearrange

$$x^2 = \dfrac{750}{120} = 6.25$$

$$x = \sqrt{6.25} = 2.5$$ — Solve for x: 1 mark

$x \propto \sqrt[3]{u}$, so $u \propto x^3$ — Make u the subject

So $u = kx^3$ — Write down the equation of proportionality: 1 mark

So $0.1 = k \times 2.5^3 = 15.625k$ — Substitute values

$$k = \dfrac{0.1}{15.625} = 0.0064$$

Calculate the constant of proportionality: 1 mark

So $u = 0.0064x^3$

When $x = 5$, $u = 0.0064x^3$

Equation of proportionality

$$= 0.0064 \times 5^3$$

Substitute for x

$$= \underline{0.8}.$$ — Calculate u: 1 mark

- The first step involves 'turning round' the proportionality equation from the first part and solving for x.

- Making u the subject of the new proportionality equation isn't absolutely necessary, but can save you a lot of time and difficult manipulation involving cube roots. As the objective is to calculate a value of u, it makes sense to have $u = kx^3$ rather than $x = k\sqrt[3]{u}$.

Specimen question 2

(a)i) Factorise the expression
$x^2 - 3x - 18$ (2 marks)

ii) Hence solve the equation
$x^2 - 3x = 18$ (1 mark)

(b) Solve the inequality $x^2 < 25$ (2 marks)

Model answer 2

(a)i) $x^2 - 3x - 18 = (x + ?)(x - ?)$

The two ?'s multiply together to make -18: hence the opposite signs: 1 mark

Possibilities are:

$(x + 1)(x - 18)$ — This will give $-17x$

$(x + 2)(x - 9)$ — This will give $-7x$

$(x + 3)(x - 6)$

This will give $-3x$: no need to look any further

$(x + 6)(x - 3)$

$(x + 9)(x - 2)$

$(x + 18)(x - 1)$

So $\underline{x^2 - 3x - 18 = (x + 3)(x - 6)}$

The complete factorisation: 1 mark

- When factorising quadratic expressions, always check that you have the correct sign in each bracket.
- Be sure to make a complete list of the factors of the number term (in this case, -18).

ii) $x^2 - 3x = 18$ can be rewritten as:

$x^2 - 3x - 18 = 0$ | Write down the equation |

So $(x + 3)(x - 6) = 0$ | From part (i) |

This is only possible if

$(x + 3) = 0$ or $(x - 6) = 0$

So $x = -3$ or $x = 6$

| The complete solution has two answers: 1 mark |

- If the original expression is equal to 0, its factorised version must be too.
- Quadratic equations that can be factorised like this have two solutions. Make sure you give them both!

(b) The equation $x^2 = 25$ has two solutions, $x = 5$ and $x = -5$.

| Find the two solutions of the equation - 1 mark |

So $x^2 < 25$ means that $x < 5$, and $x > -5$.
This can be written as a single inequality, $\underline{-5 < x < 5}$.

| Two separate inequalities would do, but this is neater - 1 mark |

- Remember that the square root of a positive number has two values: one positive, the other negative.

Specimen question 3

A sequence of numbers begins

4, 12, 26, 46, 72, ...

(a) Find an expression for the nth term of the sequence.   (3 marks)

(b) Calculate the 200th term of the sequence. (1 mark)

Model answer 3

(a)

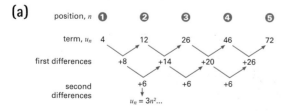

The second differences are all +6: this means that the coefficient of n^2 is 3.

| Find the number multiplying n^2: 1 mark |

Subtract $3n^2$ from each term of the sequence:

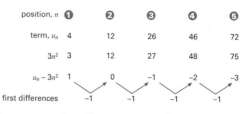

The expression for $u_n - 3n^2$ is $-n + 2$.

| Obtain the n and number terms: 1 mark |

So the nth term of the original sequence is $u_n = \underline{3n^2 - n + 2}$.

| Give the finished formula: 1 mark |

- Remember that the number in the second differences is always **double** the coefficient of n^2.
- Be careful when subtracting the $3n^2$ terms from the original sequence. It's easy to do this the 'wrong way round' and end up with $u_n = 3n^2 + n - 2$.
- Always check the first few terms of the sequence by substitution.

(b) When $n = 200$, $u_n = 3n^2 - n + 2$

| Write down the formula |

$= 3 \times 200^2 - 200 + 2$ | Substitute for n: 1 mark |

$= 120\ 000 - 198$ | Simplify |

$= \underline{119\ 802}$. | Calculate the value: 1 mark |

Further questions

1 Solve the equation $\dfrac{4x + 3}{2}$ $\dfrac{x - 10}{5} = 26$

2 Solve the simultaneous equations

$3x + 2y = 4$

$y^2 - 2x^2 = 17$

3 Make y the subject of the formula $x = \dfrac{y - 4}{1 + y}$.

4 Solve the equation $x^3 - 4x = 10$ by trial and improvement, correct to 1 decimal place.

Answers

1 $x = 12.5$

2 $x = -2, y = 5$ or $x = 26, y = -37$.

3 $y = \dfrac{x + 4}{1 - x}$.

4

x	$x^3 - 4x$	comments
1	−3	$x > 1$
2	0	$x > 2$
3	15	$x < 3$
2.5	5.625	$x > 2.5$
2.6	7.176	$x > 2.6$
2.7	8.883	$x > 2.7$
2.8	10.752	$x < 2.8$
2.75	9.796875	$x > 2.75$

$x = 2.8$ to 1 d.p.

DATA HANDLING

Data handling covers three main areas: representing data, which involves displaying data in charts and tables, and interpreting them; processing data, which includes work on averages and ranges to describe and compare frequency distributions; and probability, the study of chance events.

Specimen question 1

(a) Find the mean and standard deviation of the following set of numbers:

4, 7, 8, 8, 9, 9, 11

(b) Using your result from part (a), or otherwise, find the mean and standard deviation of the following set of numbers:

24, 27, 28, 28, 29, 29, 31

(c) Find the mean and standard deviation of the following set of numbers:

40, 70, 80, 80, 90, 90, 110

Model answer 1

(a) The mean is $\bar{x} = (4 + 7 + 8 + 8 + 9 + 9 + 11) \div 7 = 56 \div 7 = 8$.

x		4	7	8	8	9	9	11
$x - \bar{x}$		−4	−1	0	0	1	1	3
$(x - \bar{x})^2$		16	1	0	0	1	1	9

Table/list of deviations and squared deviations: 2 marks

So variance $s^2 = 28 \div 7 = 4$

Calculate mean of squared deviations

Standard deviation $s = 2$.

Take the square root of variance to calculate standard deviation: 1 mark

• This is a paper 1 question, and looks daunting at first to tackle without a calculator, but the arithmetic has been kept very simple.

(b) All the data items have been increased by 20. The mean is also increased by 20, so $\bar{x} = 28$. 1 mark

The values of $x - \bar{x}$, however, are unchanged.

Standard deviation $s = 2$. 1 mark

(c) All the data items have been multiplied by 10. The mean is also multiplied by 10, so $\bar{x} = 80$. 1 mark

The values of $x - \bar{x}$, are also multiplied by 10.

Standard deviation $s = 20$. 1 mark

Specimen question 2

In a trial of a new exam, a group of pupils were asked to finish the paper in 2 hours, but were given extra time if they needed it. This table shows the results.

Time Taken (t)	Frequency (f)
1 hr $\leqslant t <$ 1 hour 30 mins	3
1 hr 30 mins $\leqslant t <$ 1 hr 40 mins	8
1 hr 40 mins $\leqslant t <$ 1 hr 50 mins	15
1 hr 50 mins $\leqslant t <$ 2 hrs	42
2 hrs $\leqslant t <$ 2 hrs 10 mins	83
2 hrs 10 mins $\leqslant t \leqslant$ 2 hrs 30 mins	49

No pupils took less than 1 hour or longer than $2\frac{1}{2}$ hours.

(a) Construct a cumulative frequency table for this data. (2 marks)

(b) Using your table, draw a cumulative frequency graph. (3 marks)

(c) Using your graph, estimate
   i) the median time (1 mark)
   ii) the interquartile range. (2 marks)

(d) From your graph, estimate the probability that a pupil selected at random from the sample group took 2 hours and 15 minutes or more to complete the exam. (2 marks)

Model answer 2

(a)

Time taken (t)	Cumulative Frequency
$t < 1$ hr 30 mins	3
$t < 1$ hr 40 mins	11
$t < 1$ hr 50 mins	26
$t < 2$ hrs	68
$t < 2$ hrs 10 mins	151
$t \leqslant 2$ hrs 30 mins	200

2 marks

(b)

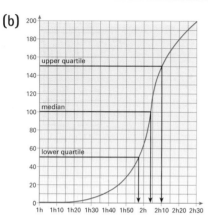

3 marks

(c) Median = $\underline{2\text{h } 3\frac{1}{2} \text{ min}}$ 1 mark

Interquartile range = 2h 10min − 1h 57min = $\underline{13\text{ min}}$ 2 marks

(d) The number of pupils corresponding to 2 hours 15 minutes on the graph is 168. This suggests that 32 pupils took this time or longer.

Find the number of pupils: 1 mark

So P(pupil took 2h 15 m or more) = $\frac{32}{200}$

$= \frac{4}{25}$ or $\underline{0.16}$.

Use the number of pupils to calculate the probability: 1 mark

- Remember to read up from the time axis to the curve for this part.

Further questions

1 A medical treatment has a 0.7 probability of success. If it doesn't work, it can be repeated, but the probability of success is reduced by 0.1 each time.

Once the probability of success drops below 50%, doctors are unwilling to prescribe the treatment.

(a) Draw a tree diagram to illustrate this situation.

(b) What is the probability that the treatment will eventually succeed?

2 In a game at a charity stall, you pay 20p to spin these two fair spinners:

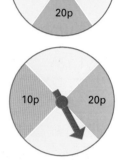

(a) Draw a possibility space diagram to show the possible outcomes.

(b) Copy and complete this table:

win	0p	10p	20p	30p	40p
probability					$\frac{1}{20}$

Answers

1 (a)

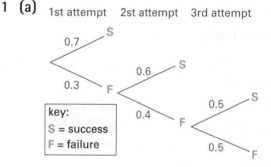

key:
S = success
F = failure

(b) 0.94

2 (a)

		1st spinner				
		0p	0p	10p	10p	20p
2nd spinner	0p	0p	0p	10p	10p	20p
	0p	0p	0p	10p	10p	20p
	10p	10p	10p	20p	20p	30p
	20p	20p	20p	30p	30p	40p

(b)

win	0p	10p	20p	30p	40p
probability	$\frac{1}{5}$	$\frac{3}{10}$	$\frac{3}{10}$	$\frac{3}{20}$	$\frac{1}{20}$

Topic checker

- Go through these questions after you've revised a group of topics, putting a tick if you know the answer, a cross if you don't.
- Try the questions again the next time you revise . . . until you've got a column that's all ticks! Then you'll know you can be confident . . .

Number

1	What is a prime number?
2	What is a prime factor?
3	Write 120 as a product of its prime factors.
4	What is the Fibonacci Sequence?
5	Evaluate x^0.
6	What is the rule of indices that is used when multiplying two powers of the same base?
7	Evaluate f^{-3}.
8	Evaluate $25^{\frac{1}{2}}$.
9	What is standard form?
10	Write the following in standard form. (a) 1 000 000 (b) 0.000 000 000 12
11	Simplify $\sqrt{12}$.

Algebra

12	Solve $6y + 10 = 52$.
13	Make x the subject of the following: (a) $scx + sx = fmp$ (b) $\sqrt{\frac{1}{x}} = y$
14	Explain the technique used to 'undo' a negative.

Answers
1 A number with only two factors and the factors are different.
2 It is a factor that is itself prime.
3 $2^3 \times 3 \times 5$
4 A sequence of numbers where the subsequent number is found by adding the two previous numbers,
e.g. 1, 1, 2, 3, 5, is made up from $0 + 1 = 1, 1 + 1 = 2, 1 + 2 = 3, 2 + 3 = 5$.
5 1
6 The indices are added.
7 $\frac{1}{f^3}$
8 5
9 A way of writing very large or very small numbers and making them more managable.
10 (a) 1×10^6 (b) 1.2×10^{-10}
11 $2\sqrt{3}$
12 $y = 7$
13 (a) $x = \frac{fmp}{sc + s}$ (b) $x = \frac{1}{y^2}$
14 Multiply through by negative one.

15 Solve the inequality $3y - 12 \geq 18$.

16 What is the gradient and y-intercept of the graph of $3x + y = 5$?

17 How do you find the equation of a line parallel to a given line, if you know its y-intercept?

18 Solve the simultaneous equations:
$2x + y = 13$
$5x + 2y = 29$.

19 What is the highest common factor of $6x^2 + 36x$?

20 What is the lowest common multiple of $5x$ and $6x$?

21 Factorise $20y^4 - 6y^2$.

22 Expand and simplify $(x + 3)(x + 9)$.

23 Expand $(x - 8)^2$.

24 What is a difference of two squares and why is it special?

25 For a given sequence $u_n = 4n + 5$, find the fourth term in the sequence.

26 Find a formula for this sequence: 6, 11, 16, 21, 26, ...

27 Rearrange $V = \sqrt{\frac{t}{\pi}}$ to make t the subject.

28 What is a quadratic equation?

29 What is the reciprocal of $\frac{1}{7}$?

30 What is a cubic equation?

31 Factorise $x^2 - 9$.

32 By using trial and improvement, solve $x^3 + 2x^2 + 2 = 177$.

33 Simplify $(a^m)^n$.

34 Simplify $d^6 \times d^5$.

35 Simplify $2t^6 \div t^3$.

36 Simplify $(3y^2)^5$.

Answers

15 $y \geq 10$
16 gradient $= -3$, intercept $= 5$
17 Since the line is parallel to a known line, it must have the same gradient and, therefore, must fit the $y = mx + c$ general form.
18 $x = 3$, $y = 7$
19 $6x$
20 $30x$
21 $2y^2(10y^2 - 3)$

22 $x^2 + 12x + 27$
23 $x^2 - 16x + 64$
24 It is of the general form $x^2 - y^2$ and is a result of the expansion $(x + y)(x - y)$. The special feature is that in the expansion, the middle terms cancel each other out.
25 21
26 $5x + 1$
27 $t = V^2\pi$
28 It is an equation of whole non-negative

powers whose highest power is 2.
29 7
30 It is an equation of whole non-negative powers whose highest power is 3.
31 $(x + 3)(x - 3)$
32 $x = 5$
33 a^{mn}
34 d^{11}
35 $2t^3$
36 $243y^{10}$

Algebra

37 Simplify $6(h^2)^4$.

38 Simplify $25f^2g^3h^6 \div 5tg^2$.

39 Simplify $12mn^3 \times 3mnp^2$.

40 Solve for k, the equation $z^k = 1$.

41 Solve $h^{\frac{1}{2}} = 25$.

42 Factorise $x^2 - 8x$.

43 Factorise $169z^2 - 225y^2$.

44 Simplify $\frac{x^3y^2z}{z} \times \frac{x^4yz}{z^3}$.

45 Simplify $6x \div \frac{1}{x}$.

46 Simplify $3(x + 6) \times 4(x + 7)$.

47 Work out $\frac{5(x + 4)}{5x^2}$.

48 Simplify $12xy^2 \div \frac{2}{xy}$.

49 Solve $25h^2 = 100$.

50 Solve $m^2 + 3m - 12 = -2$.

51 What is the formula for solving the quadratic equation $ax^2 + bx + c = 0$?

52 Solve the simultaneous equations:
$x^2 + y^2 = 20$
$2x - y = 0$.

Shape, space and measures

53 What are the tests for congruence?

54 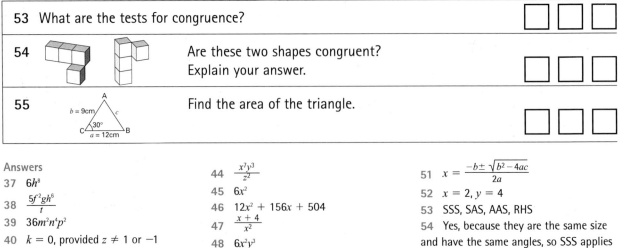 Are these two shapes congruent? Explain your answer.

55 Find the area of the triangle.

$b = 9\text{cm}$, A, c, $30°$, C, $a = 12\text{cm}$, B

Answers

37 $6h^8$

38 $\frac{5f^2gh^6}{t}$

39 $36m^2n^4p^2$

40 $k = 0$, provided $z \neq 1$ or -1

41 $h = 625$

42 $x(x - 8)$

43 $(13z - 15y)(13z + 15y)$

44 $\frac{x^7y^3}{z^2}$

45 $6x^2$

46 $12x^2 + 156x + 504$

47 $\frac{x + 4}{x^2}$

48 $6x^2y^3$

49 $h = 2$

50 $m = -5$ or 2

51 $x = \frac{-b \pm \sqrt{b^2 - 4ac}}{2a}$

52 $x = 2, y = 4$

53 SSS, SAS, AAS, RHS

54 Yes, because they are the same size and have the same angles, so SSS applies and SAS applies.

55 27cm^2

56 Write down the sine rule.

57 Write down the cosine rule.

58 Find x.

59 What is a translation?

60 What happens when a shape is enlarged by a negative scale factor?

61 If $f(x) = x^3 + 1$, find:
   (a) $f(4)$ **(b)** $f(9)$.

62 If $f(x) = 3x^2$, find $f(5x)$.

63 Given $p = \begin{pmatrix} 6 \\ 5 \end{pmatrix}$ and $q = \begin{pmatrix} 3 \\ 5 \end{pmatrix}$, find $p + q$.

64 Given $r = \begin{pmatrix} 5 \\ 9 \end{pmatrix}$, find $5r$.

65 Given $m = \begin{pmatrix} 8 \\ -3 \end{pmatrix}$ and $n = \begin{pmatrix} -2 \\ -6 \end{pmatrix}$, find:

   (a) $m - n$ **(b)** $3m$ **(c)** $3m - 2n$.

66 What is the acute angle with the same sine as sin 150°?

67 What is the acute angle whose cosine is the same as sin 30°?

68 What is the equation of the line that is parallel to $y = 2x + 1$ and passes through the point (0, 4)?

69 A line joins the points (1, 2) and (3, 7). What is the equation of the line?

70 What is the formula to find the area of a trapezium?

Circle geometry

71 Can you draw a tangent to a circle?

72 What is the size of the angle in a semi-circle?

Answers

56 $\dfrac{a}{\sin A} = \dfrac{b}{\sin B} = \dfrac{c}{\sin C}$

57 $a^2 = b^2 + c^2 - 2bc \cos A$

58 12.75 cm

59 It is a shift in the plane, e.g. $\begin{pmatrix} 4 \\ 3 \end{pmatrix}$ will translate a point 4 units in the positive x direction and 3 units in the positive y direction.

60 The measurements are in the opposite direction.

61 (a) 65   (b) 730

62 $75x^2$

63 $\begin{pmatrix} 9 \\ 10 \end{pmatrix}$

64 $\begin{pmatrix} 25 \\ 45 \end{pmatrix}$

65 (a) $m - n = \begin{pmatrix} 10 \\ 3 \end{pmatrix}$ (b) $3m = \begin{pmatrix} 24 \\ -9 \end{pmatrix}$

     (c) $3m - 2n = \begin{pmatrix} 28 \\ 3 \end{pmatrix}$

66 30°

67 60° because cos 60 = sin 30 = 0.5

68 $y = 2x + 4$

69 $y = 2.5x - 0.5$

70 $A = \frac{1}{2}(a + b)h$

71 Student by student.

72 90°

Circle geometry

73 What can you say about the angle subtended by an arc at the centre of a circle, in comparison to the angle subtended by the same arc at any part of the remaining circumference?

74 What can you say about angles in the same segment?

75 How do I draw a perpendicular bisector to a line?

76 What is the difference between a segment of a circle and a sector of a circle?

77 What is the formula for the volume of a sphere?

Data handling

78 What is special about a histogram?

79 What is frequency density?

80 What is the variance?

81 What is the standard deviation of a set of data?

82 Find the standard deviation for these sets of data.
(a) 1   5   9   9
(b) 2   7   7   8   9
(c) 5   6   9   9   11

83 What is the mean and standard deviation of:
€25, €46, €56, €80, €92?

Probability

84 What is the OR rule in probability?

85 A bag contains 5 blue balls, 4 black balls and 8 white balls. What is the probability of picking out a blue or a white ball?

Answers

73 The angle at the centre is twice as big as the angle subtended by the same arc at any point on the remaining part of the circumference.

74 They are equal.

75 (i) Open up a pair of compasses to more than half of the line length.
(ii) Draw arcs on both sides of the line segment by placing the pair of compasses at one end of the line segment. (iii) Now put the pair of compasses at the other end of the line segment and repeat. (iv) You should now have two pairs of arcs, one each side of the line, that cross. Join the two crossing points with a straight line – this is the perpendicular bisector.

76 A sector is bounded by part of the circumference and two radii; a segment is bounded by part of the circumference and a straight line.

77 $V = \frac{4}{3}\pi r^3$

78 The frequency is proportional to the area under the bar.

79 It is a measure of the density of the data in class intervals.

80 This is the mean of the square of the deviations of the data from the mean.

81 This is the square root of the variance.

82 (a) st dev = 3.32 (b) st dev = 2.42 (c) st dev = 2.19

83 mean = €59.80, standard deviation = €23.92

84 The OR rule occurs when events have mutually exclusive outcomes. Mathematically, you write this as $P(X \text{ or } Y) = P(X) + P(Y)$.

85 $\frac{8}{17}$

86 A playing card is selected from a fair pack of 52 playing cards.
What is the probability that it is not an Ace?

87 What is the AND rule in probability?

88 When a fair coin is spun through the air and a playing card is drawn from a fair pack, what is the probability of getting:
(a) a head
(b) a black card
(c) a black card and a head?

89 A bag contains 10 balls, 6 are red and 4 are white.
Two balls are picked from the bag and NOT replaced,
what is the probability that 1 red and 1 white ball are selected?
Draw a tree diagram to show this information.

90 In a pet shop there are two pens of guinea pigs.
In pen 1 there are 14 guinea pigs, 6 of them are white and the rest are brown.
In pen 2 there are 18 guinea pigs, 12 are brown and the rest are white.
A pig is chosen from each pen.
Use a tree diagram to work out the probability that both pigs are brown.

91 What is the definition of mutually exclusive outcomes of an event?

92 What are independent events?

93 If you know the probability of an outcome occurring, how do you calculate
the probability of it not occurring?

94 What is the probability of a certainty?

95 What is the probability of impossibility?

96 If the probability that it will not rain in St Helier is 0.7, do I need to take an umbrella?

97 How is relative frequency defined?

Answers

86 $\frac{12}{13}$

87 The AND rule occurs where events have independent outcomes, for instance, rolling a dice and spinning a coin through the air. Mathematically, you write this as $P(X \text{ and } Y) = P(X) \times P(Y)$.

88 **(a)** $\frac{1}{2}$ **(b)** $\frac{1}{2}$ **(c)** $\frac{1}{4}$

89
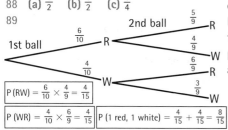
$P(RW) = \frac{6}{10} \times \frac{4}{9} = \frac{4}{15}$
$P(WR) = \frac{4}{10} \times \frac{6}{9} = \frac{4}{15}$ $P(\text{1 red, 1 white}) = \frac{4}{15} + \frac{4}{15} = \frac{8}{15}$

90 $BB = \frac{8}{21}$

91 Where the outcome of one event means that an alternative event cannot happen. For example, when a coin is spun and a head occurs, it means a tail cannot occur.

92 Independent events are events whose outcomes are not dependent on each other. For example, lottery balls this week are totally independent of previous performance, therefore, it is statistically as likely that they will reoccur as not.

93 $P(\text{outcome not occuring}) = 1 - P$ (of the outcome occuring)

94 1

95 0

96 The probability of it not raining is high, therefore, it should not be necessary to take an umbrella.

97 This is the measure of probability based on the past experience of the event, e.g. the probability of an earthquake.

A Number: fractions

1 $\frac{1}{2}$ of $\frac{1}{3}$? $\frac{1}{3}$ of $\frac{1}{2}$ Enter a mathematical sign from this list, $+, -, =, >, \leqslant,$ instead of the ?, to make this a true statement. _____

2 $4\frac{1}{3}$ as an improper fraction is _____

3 $\frac{19}{4}$ as a mixed number is _____

4 Calculate $4\frac{1}{3} \times 5\frac{5}{9}$ _____

5 Calculate $5\frac{3}{4} \div 3\frac{1}{2}$ _____

B Number patterns

1 The first four square numbers are ___, ___, ___, ___.

2 The next cube number after 64 is _____

3 120 expressed as a product of its prime factors is _____

4 The next prime number after 13 is _____

5 The fourth triangular number is _____

C Indices

1 Copy this statement, replacing stars by suitable numbers:
$(4^*)^2 = 4^{(2 \times *)} = 4^6$ _____

2 $x^k = y$. What is y when $k = 0$? _____

3 $\dfrac{a^m}{a^n} =$ _____

4 $9^h = 3$ means that $h =$ _____

5 The cube root of 27 is _____

D Money and percentages

1 On the day we checked, €1 = 60p. At that time, €50 = £ _____

2 €1500 increased by 30% = € _____

3 A coat is reduced in a sale by 25%. The original price was £120. This means the sale price is _____

4 When £40 is increased by 30%, it becomes _____

5 When £60 is decreased by 30% the amount becomes _____

E   Surds

1   A surd is _____

    _____

2   $\sqrt{2}$ is a surd because _____

    _____

3   $\sqrt{12} =$ _____

4   $2\sqrt{3} + 5\sqrt{3} =$ _____

5   $\sqrt{18} + \sqrt{24} =$ _____

F   Algebra

1   Solve the equation:
    $5x - 6 = 4x + 3$ _____

2   Solve the equation:
    $4x + 6 = 10x - 24$ _____

3   Solve the equation:
    $-\frac{1}{3}x = 42$ _____

4   Copy and complete, to make x the subject of the formula:
    $\frac{mx}{p} = c$ _____

5   Copy and complete, making x the subject of the formula:
    $6 - fx = 5x + h$

    _____

6   Copy and complete, to solve the inequality.

    $4x + 6 > 34$ _____

7   What is the reciprocal of x? _____

8   Copy and complete the factorisation of $x^2 + 9x = x($ $)$ _____

9   Copy and complete $(a + b)(a - b) =$ _____

10 Factorise $9z - 15z^2$ _____

11 $225y^2 - 169z^2$ is a special factorisation and factorises as _____

12 Complete the equation:

    $\dfrac{x^6}{y^2} \times \dfrac{x^3}{x^2} = \underline{\quad}$ _____

F   Algebra (continued)

13   Complete the equation: $\dfrac{6}{x} \div \dfrac{1}{x} = $ _____

14   Solve the equation: $x^2 = 25x$ _____

15   Solve the equation: $m^2 + 20m = -36$

_____

16   Solve the equation: $(x + 4)^2 = 121$ _____

17   The word quadratic means _____

_____

18   The word cubic means _____

_____

19   The word coefficient means _____

_____

G   Shape: congruence and trigonometry

1   The tests for congruence of triangles are _____

2   A rectangle of dimensions 7 cm by 14 cm is enlarged to 14 cm by 28 cm.
This means the scale factor of enlargement is _____

3   The area of a triangle with two sides a and b and included angle C can be
calculated by using trigonometry. Here we use the formula $A = $ _____
Complete the formula.

4   The sine rule is _____

5   The cosine rule is _____

H   Transformations

1   A translation is a _____

2   A reflection occurs when an object is reflected in a _____line and

creates an _____. Every point on the image is an_____distance

from the image line, compared with the equivalent point on the original object.

3   The location about which a shape can be rotated, is a point in the plane and is

called the _____

4   When an enlargement occurs by using a scale factor of less than one, it means

_____

5   When an enlargement occurs by using a negative scale factor, it means

_____

_____

I   Functions

1   A function is defined as _____

_____

2   If the function f is 'square', then for the input x, the output is _____

3   When $f(x) = x^3$, $f(x + 2) =$ _____

4   The graph of $y = x^2 + 3$ is a translation of $y = x^2$ in the_____ direction

along the axis by _____units.

5   When $y = x^2$ is translated 5 units in the positive direction along the x-axis, the

new equation is _____

J   Vectors

1 (a) To find the magnitude of the vector $\overrightarrow{AB} = \begin{pmatrix} 7 \\ 4 \end{pmatrix}$ use the rule of _____

(b) Complete the calculation of the magnitude_____

2   Velocity is the rate of change of _____

with respect to _____

3   On a velocity-time graph, the gradient at a point tells us the _____
of the object at the point.

4    $\begin{pmatrix} 3 \\ 2 \end{pmatrix} + \begin{pmatrix} ? \end{pmatrix} = \begin{pmatrix} -2 \\ 4 \end{pmatrix}$

5   When $r = \begin{pmatrix} 5 \\ -3 \end{pmatrix}$ $5r = \begin{pmatrix} ? \end{pmatrix}$

6   Scalar means _____

7   Magnitude means _____

8   Displacement means _____

K   Circles, cylinders and cones

1   The perpendicular bisector of a chord of a circle passes through the _____

2   The angle in a semi-circle is _____

K   Circles, cylinders and cones (continued)

3   Angles in the same segment are _____

4   The angle subtended by an arc of a circle is _____

_____

5   Allied angles are _____

6   A cyclic quadrilateral is _____

_____

7   The angle between a tangent and a chord is equal to _____

_____

8   Finding an arc length of a circle is straightforward when you use the formula arc

length: _____

9 (a) The formula for the surface area of a cylinder is _____

   (b) A cone is defined as a _____

10   To find the volume of a cylinder of radius r and height h, you use the formula

_____

L   Data handling

1   In a histogram, the frequencies of the collected data are proportional to _____

_____

2   The frequency density is defined as _____

3   The dispersion of a set of data is a measurement of _____

4   The range of a set of data is _____

_____

5   The median of a set of data is _____

_____

6   The mean of a set of data is _____

_____

7   The mode of a set of data is _____

_____

8   The interquartile range is a measure of dispersion and is defined as _____

_____

9   The deviation from the mean is a phrase that explains _____

L   Data handling (continued)

10 (a) The mean of the deviations from the mean is always _____ because

_____

(b) The mean of the square of the deviations from the mean.

This is called the _____

(c) The standard deviation of a set of data is _____

_____

M   Probability

1   If two events have mutually exclusive outcomes, it means _____

_____

2   When a coin is spun through the air, the outcomes are _____

3   A quick way of working out the probability that something will NOT happen, when you know the probability that it will happen is to calculate _____

_____

4   If X and Y are mutually exclusive outcomes of an event, $P(X$ or $Y) =$ _____

5   When one event does not depend on another event occurring, the events are called _____

6   Relative frequency is a term used to describe the probability of an outcome

based on _____

7   An event that is impossible has a probability of _____

8   An event that is certain has a probability of _____

9   If X and Y are outcomes of two independent events, $P(X$ and $Y) =$ _____

10   When a card is drawn from a fair pack, and not replaced, and then another card is drawn, the probability that they are both clubs is _____

11   A bag contains lottery balls. There are 50 balls, numbered 1 to 50.

The probability that the first ball picked is NOT 25 is _____

12   Imagine a bag containing 15 counters, 12 are red and 3 are other colours (not red).

i)   If two counters are picked out one after the other at random and the first is not replaced before the second pick occurs, then the probability that they will both be red counters is _____

ii)   If the first counter is replaced after it is picked and before the second counter is picked, then the probability of both counters being red is now _____

Answers

Fractions

1 $\frac{1}{2}$ of $\frac{1}{3}$ = $\frac{1}{3}$ of $\frac{1}{2}$
2 $\frac{13}{3}$
3 $4\frac{3}{4}$
4 $24\frac{2}{27}$
5 $1\frac{9}{14}$

Number patterns

1 1, 4, 16, 25
2 125
3 $2^3 \times 3 \times 5$
4 17
5 10

Indices

1 $(4^3)^2 = 4^{2\times3} = 4^6$
2 1
3 a^{m-n}
4 $\frac{1}{2}$
5 3

Money and percentages

1 £30
2 €1950
3 £90
4 £52
5 £42

Surds

1 a mechanism for writing an irrational number as an exact quantity. For example, $\sqrt{2}$ is an exact value, whereas 1.41 is an approximation to the square root of 2.
2 it is an exact quantity that cannot be totally determined, see above.
3 $2\sqrt{3}$
4 $7\sqrt{3}$
5 $3\sqrt{2} + 2\sqrt{6}$

Algebra

1 $x = 9$
2 $x = 5$
3 $x = -126$
4 $x = \frac{pc}{m}$
5 $x = \frac{6-h}{5+f}$
6 $x > 7$
7 $\frac{1}{x}$
8 $x(x + 9)$
9 $a^2 - b^2$
10 $3z(3 - 5z)$
11 $(15y + 13z)(15y - 13z)$
12 $\frac{x^7}{y^2}$
13 6
14 $x = 25$ or 0
15 $m = -18$ or -2
16 $x = 7$
17 of the second degree, so a quadratic equation has the highest power of 2.
18 of the third degree, so a cubic equation has the highest power of 3.
19 the numerical part of a term, usually written before the literal part, for example, in $2x$, the coefficient of x is 2.

Shape: congruence and trigonometry

1 RHS, SAS, AAS, SSS
2 2
3 $ab \sin C$
4 $\frac{a}{\sin A} = \frac{b}{\sin B} = \frac{c}{\sin C}$
5 $a^2 = b^2 + c^2 + 2bc \cos A$

Transformations

1 is a shift in the plane.
2 mirror   image   equal
3 centre of rotation.
4 the shape is reduced in size.
5 the enlargement is measured from the centre of the enlargement, in the opposite direction.

Functions

1 a relationship that associates exactly one object from one set, with each object from another set.
2 x^2
3 $(x + 2)^3$
4 positive   3
5 $y = (x - 5)^2$

Vectors

1 (a) Pythagoras

   (b) $x^2 = 7^2 + 4^2$
   $= 49 + 16$
   $x^2 = 65$
   $x = 8.06$ cm

2 distance time.
3 acceleration.
4 $\begin{pmatrix} -5 \\ 2 \end{pmatrix}$
5 $\begin{pmatrix} 25 \\ -15 \end{pmatrix}$
6 the numerical quantity does not have direction.
7 size.
8 distance covered in a direction.

Circles, cylinders and cones

1 centre.
2 a right angle.
3 are equal.
4 twice the angle subtended by the same arc at any point on the remaining part of the circumference.
5 supplementary.
6 a quadrilateral where all of the four corners of its four

sides are inscribed in the circumference of the circle.
7 the angle in the alternate segment.
8 $\frac{\pi r \theta}{180°}$
9 (a) $s = 2\pi rh + 2\pi r^2$
(b) as a pyramid with a circular base.
10 $V = \pi r^2 h$

7 0
8 1
9 $P(x) \times P(Y)$
10 $\frac{3}{51}$
11 $\frac{49}{50}$
12 i) $\frac{12}{15} \times \frac{11}{14} = \frac{22}{35}$
ii) $\left(\frac{12}{15}\right)^2 = \frac{16}{25}$

Data handling

1 area under the bar.
2 $\frac{frequency}{class\ width}$
3 the spread of the data.
4 highest value to the lowest value and it is a single value.
5 the middle value when the data is in order.
6 the arithmetic average, where the data is summed and the resulting answer is divided by the number of items.
7 the item of data that occurs the most often.
8 as the middle 50% of the spread of data.
9 the distance a value is away from the mean.
10 (a) zero the positive and negative values cancel each other out.
(b) variance
(c) the square root of the variance.

Probability

1 one outcome precludes the occurrence of the other outcome.
2 mutually exclusive.
3 the probability that it will happen and subtract from 1.
4 $P(X) + P(Y)$
5 independent events.
6 previous experience.

Answers

Fractions and decimals (p 9)

1 (a) terminate (b) recur
   (c) terminate (d) terminate
   (e) recur   (f) recur
   (g) terminate (h) terminate
   (i) recur   (j) terminate

2 (a) 0.7   (b) 0.21
   (c) 0.88   (d) 0.8$\dot{3}$
   (e) 0.$\dot{2}\dot{7}$ (f) 0.11$\dot{6}$

3 (a) $\frac{11}{25}$ (b) $\frac{19}{2000}$
   (c) $\frac{1}{16}$ (d) $\frac{1}{9}$
   (e) $\frac{107}{333}$ (f) $\frac{112}{495}$

Working with indices (p 11)

1 (a) 3   (b) $\frac{1}{3}$ (c) 27
2 (a) 1   (b) 5^{-6}, 0.000064
   (c) 8   (d) $\frac{1}{8}$
3 8
4 1
5 1

Numbers in standard form (p 13)

1 (a) 6×10^2 (b) 2×10^6
   (c) 4.1×10^4 (d) 9.55×10^5
   (e) 9×10^{-4} (f) 3×10^{-3}
   (g) 1.01×10^{-6} (h) 5.8×10^{-1}
2 (a) 8000   (b) 400 000
   (c) 2 100 000 000 (d) 70 030 000
   (e) 0.3   (f) 0.00000005
   (g) 0.0575   (h) 0.00000677
3 2.65×10^{52} 4 (a) 6×10^{17} (b) 1.5×10^{-7}
5 (a) 1.333×10^9 (b) 1.369×10^{-11}
6 0.15

Ratio, proportion and percentages (p 15)

1 (a) $1:125$ (b) $1:8$
2 (a) € 56.45   (b) € 229.03
   (c) € 2419.35
3 4970   4 $666\frac{2}{3}$ cm   5 £63

Surds (p 17)

1 $4\sqrt{2}$ 2 $10\sqrt{3}$ 3 $7\sqrt{5}$
4 $\frac{\sqrt{3}}{4}$ 5 $\frac{2\sqrt{2}}{3}$ 6 $\sqrt{\frac{2}{100}} = \frac{\sqrt{2}}{10}$
7 $3\sqrt{7} + \sqrt{21}$ 8 $\sqrt{2} - 2$ 9 $4\sqrt{10} + 30$
10 $8 + 3\sqrt{6}$ 11 $6 - 5\sqrt{3}$ 12 $60 - 52\sqrt{5}$
13 $\frac{3\sqrt{11}}{11}$ 14 $\frac{\sqrt{6}}{15}$ 15 $\frac{1}{\sqrt{3}}$
16 $3 + 2\sqrt{3}$ 17 $-\frac{3 + \sqrt{5}}{2}$

Solving simple equations (p 19)

1 $x = 7$ 2 $y = 9$ 3 $d = 2$ 4 $f = 12$
5 $c = 5$ 6 $h = \frac{1}{2}$ 7 $x = 3$ 8 $x = -1$
9 $k = 6$ 10 $p = \frac{2}{3}$

Rearranging formulae (p 21)

1 $x = r - 9$ 2 $x = a + z$
3 $x = \frac{16 - 4y}{5}$ 4 $x = \sqrt{\frac{c}{m}}$
5 $x = 4m$ 6 $x = p^2 + pc$
7 $x = \frac{cb}{m}$ 8 $x = \frac{5-p}{3+f}$
9 $y = \frac{9-2x}{5}$ 10 $y = \frac{x}{2} - 5$

Inequalities (p 23)

1 $x \leqslant 5$ 2 $x > -3\frac{1}{2}$ 3 $a \geqslant \frac{1}{6}$
4 $t \geqslant -1\frac{1}{3}$ 5 $d > -5$ 6 $w \leqslant 2\frac{1}{2}$
7 $f \geqslant -8$ 8 $\frac{3}{2} > p > \frac{1}{2}$ 9 $-\frac{1}{5} \leqslant r \leqslant 2$
10 $-15 < t < 9$

Lines and equations (p 25)

1 (a) gradient = 7, y-intercept = 3
   (b) gradient = 3, y-intercept = -5
   (c) gradient = 1, y-intercept = 12
   (d) gradient = 3, y-intercept = -2
2 (a) gradient = -3, y-intercept = 4
   (b) gradient = 2, y-intercept = -7
   (c) gradient = $\frac{1}{3}$, y-intercept = $-\frac{7}{3}$
   (d) gradient = $-\frac{1}{5}$, y-intercept = $\frac{9}{5}$
3 $y = 4x + 5$ 4 $y = x + 6$ 5 $y = 2x + 1$
6 Point of intersection = (3, 2). Equation is $y = \frac{1}{2}x + \frac{1}{2}$.

Simultaneous equations and graphs (p 27)

1 $x = 3$, $y = 7$

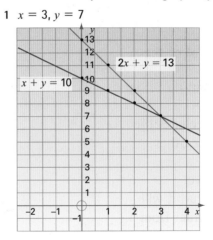

2 $c = 15 + 5d$ and $c = 7.50d$;
  $c = £45$, $d = 6$

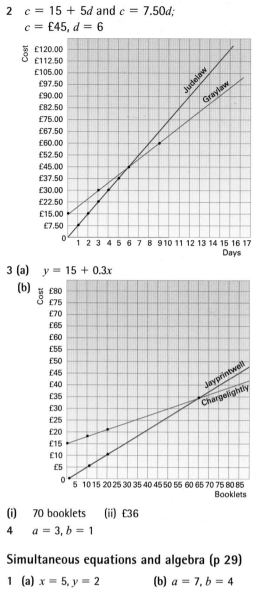

3 (a) $y = 15 + 0.3x$

(b)

(i) 70 booklets   **(ii)** £36

4 $a = 3$, $b = 1$

Simultaneous equations and algebra (p 29)

1 (a) $x = 5$, $y = 2$ **(b)** $a = 7$, $b = 4$
  (c) $m = 9$, $n = 1$ **(d)** $t = 6$, $r = 7$
  (e) $x = 5$, $y = 4$ **(f)** $c = 7$, $d = 9$
  (g) $e = 12$, $f = 7$ **(h)** $p = 15$, $q = -2$
  (i) $a = 2$, $b = 1$ **(j)** $p = 4$, $q = 2$

2 $a = 60$, $b = 8$

3 (a) $a = -\frac{2}{3}$ $b = -\frac{8}{9}$ **(b)** $m = \frac{3}{4}$

Using brackets in algebra (p 31)

1 $x = 4$ **2** $x = 10$

3 $x = -8$ **4** $x = 4.5$

5 $x = -5\frac{1}{3}$ **6** $x = 4$

7 $x = 15\frac{1}{2}$

8 (a) $7x(2x + 1)$ **(b)** $9y(4y - 1)$

9 (a) $5y^2(3y^2 + 5)$ **(b)** $20a(5a + b^3)$

Multiplying bracketed expressions (p 33)

1 $x^2 + 4x + 3$ **2** $x^2 + 9x + 14$

3 $x^2 + 4x + 4$ **4** $x^2 + nx + mx + mn$

5 $x^2 + 6x + 8$ **6** $x^2 - 2ax + a^2$

7 $x^2 - 6x + 9$ **8** $a^2 - 2ax + x^2$

9 $x^2 - 2xy + y^2$ **10** $x^2 - 4$

11 $x^2 - 9y^2$ **12** $36 - x^2$

Regions (p 35)

1

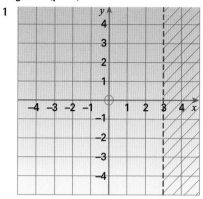

2

3

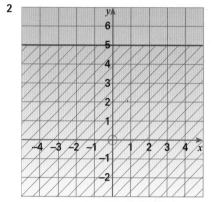

4

5

6

7

8

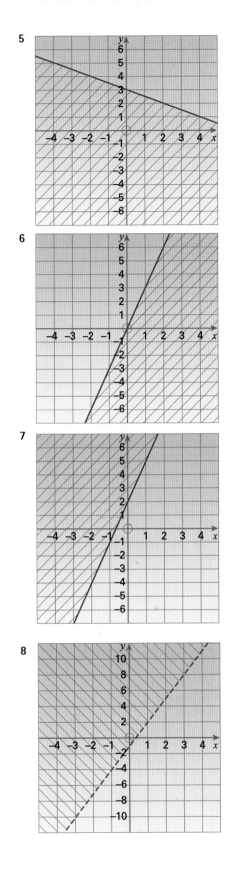

9

10

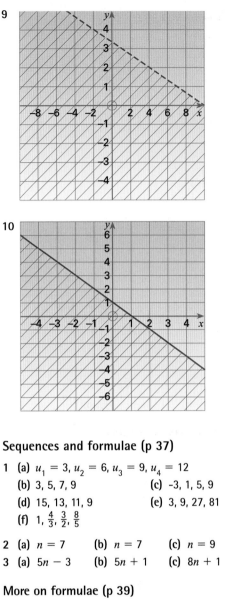

Sequences and formulae (p 37)

1 (a) $u_1 = 3$, $u_2 = 6$, $u_3 = 9$, $u_4 = 12$
   (b) 3, 5, 7, 9   (c) -3, 1, 5, 9
   (d) 15, 13, 11, 9   (e) 3, 9, 27, 81
   (f) 1, $\frac{4}{3}$, $\frac{3}{2}$, $\frac{8}{5}$

2 (a) $n = 7$ (b) $n = 7$ (c) $n = 9$
3 (a) $5n - 3$ (b) $5n + 1$ (c) $8n + 1$ (d) $3n - 1$

More on formulae (p 39)

1 22.00	2 196.07 cm²	3 29.22 cm
4 1960.36	5 5026.55	6 6.95
7 4.47	8 6.41	

Equations of proportionality (p 41)

1 $y = \dfrac{3x^2}{512}$

speed (x km/h)	41.3 (1 dp)	64	80
braking distance (y metres)	10	24	37.5

2 $v = \dfrac{5\sqrt{30}}{\sqrt{d}}$

planet	Mercury	Earth	Neptune
orbital distance (d au)	0.33 (2 sf)	1	30
speed (v km/s)	48	27 (2 sf)	5

Quadratic functions (p 43)

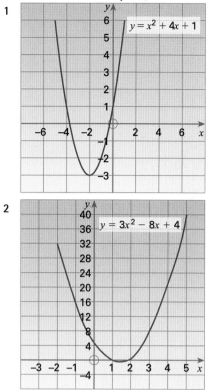

1 $y = x^2 + 4x + 1$

2 $y = 3x^2 - 8x + 4$

3 The graph meets the y-axis at the origin.

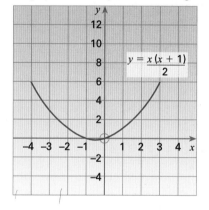

$y = \dfrac{x(x+1)}{2}$

Cubic functions (p 45)

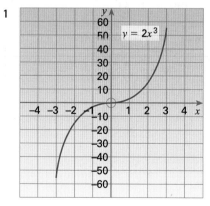

1 $y = 2x^3$

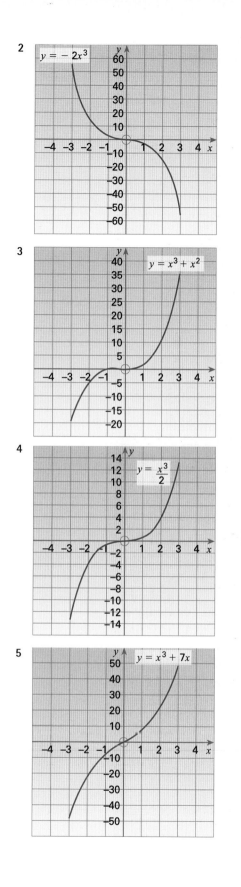

2 $y = -2x^3$

3 $y = x^3 + x^2$

4 $y = \dfrac{x^3}{2}$

5 $y = x^3 + 7x$

6

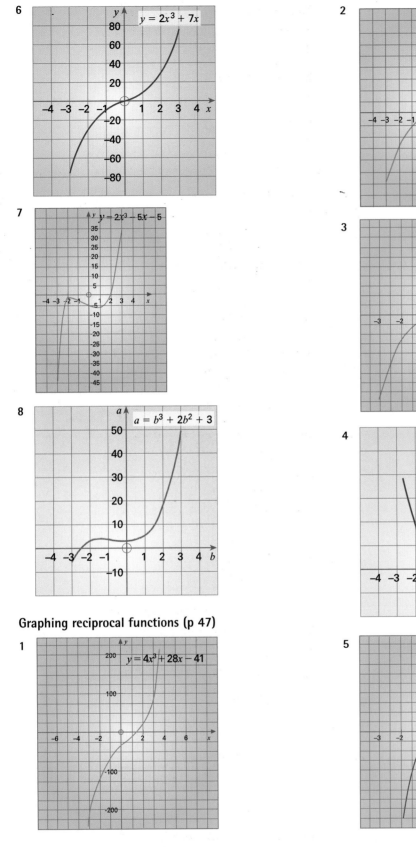

$y = 2x^3 + 7x$

7

$y = 2x^3 - 5x - 5$

8

$a = b^3 + 2b^2 + 3$

Graphing reciprocal functions (p 47)

1

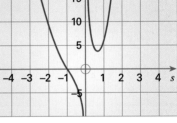

$y = 4x^3 + 28x - 41$

2

$y = m - \dfrac{m^3}{10}$

3

$y = 3x^3 + 2x$

4

$t = 4s^2 + 3s + \dfrac{1}{s}$

5

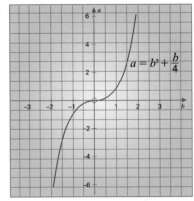

$a = b^3 + \dfrac{b}{4}$

6

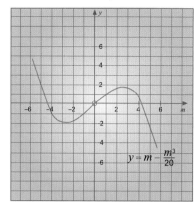

$y = m - \dfrac{m^3}{20}$

Solving equations with graphs (p 49)

1 Solutions are $x = +2$ or $x = -2$

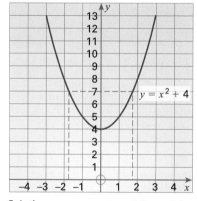

2 Solutions are $x = +1.7$ or -1.7

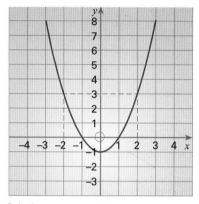

$y = x^2 + 4$

3 Solutions are $x = $ approx -1.9 or $+3.9$

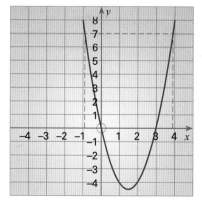

Trial and improvement (p 51)

1 1.6	**2** 2.1	**3** 3.5
4 1.8	**5** 1.7	

More complicated indices (p 53)

1 d^9	**2** $12r^{13}$	**3** g^5
4 $30f^3 g^4$	**5** x^7	**6** t^4
7 $8m^2$	**8** $\dfrac{1}{x}$	**9** $\dfrac{6}{j^2}$
10 $4y^6$	**11** $a^3 b^{12}$	**12** $25x$

Indices and equations (p 55)

1 $y = \dfrac{1}{2}$	**2** $k = 0$	**3** $y = 3$	**4** $x = 625$
5 $k = 3$	**6** $k = 1$	**7** $y = -7$	

Factorising quadratic expressions (p 57)

1 (a) $x(x + 4)$ (b) $9(x + 2)$
  (c) $(x + 3)(x + 12)$
2 (a) $x(x + 6)$ (b) $12(x - 12)$
  (c) $(x + 5)(x - 2)$
3 (a) $(x - 4)(x - 4)$ (b) $2(x - 2)(x - 1)$
  (c) $6(x - 3)(x - 4)$
4 (a) $10(x + 1)(x + 1)$ (b) $8(a + 1)(a + 1)$
  (c) $(m - 3)(m - 3)$

The difference of two squares (p 59)

1 $(x + y)(x - y)$
2 Not a difference of two squares.
3 $(7t - 8m)(7t + 8m)$
4 Not a difference of two squares.
5 Not a difference of two squares.
6 Not a difference of two squares.
7 $(10k + 9m)(10k - 9m)$
8 $(13c + 15d)(13c - 15d)$
9 $625k^2 - 400m^2 = 25\,(5k - 4m)(5k + 4m)$
10 $1600v^2 - 400u^2 = 400\,(2v - u)(2v + u)$
11 $18p^2 - 2q^2 = 2\,(3p - q)(3p + q)$
12 $k^2 - \dfrac{m^2}{4} = (k - \dfrac{m}{2})(k + \dfrac{m}{2})$
13 $a^4 - b^4 = (a^2 - b^2)(a^2 + b^2) = (a - b)(a + b)(a^2 + b^2)$
14 $G^2 - \dfrac{1}{H^2} = (G - \dfrac{1}{H})(G + \dfrac{1}{H})$
15 $\dfrac{a^2}{3} - \dfrac{1}{48b^6} = \dfrac{1}{12}(2a - \dfrac{1}{2b^3})(2a + \dfrac{1}{2b^3})$

Algebraic fractions (p 61)

1 $\dfrac{y^3}{z}$	**2** $1\dfrac{1}{5}$	**3** $\dfrac{1}{2x - 8}$	**4** $5x^2$
5 $4m^2 x^3$	**6** $\dfrac{6a + 7}{20}$	**7** $\dfrac{5t + 8}{6}$	

Quadratic equations 1 (p 63)

1 $x = 0$ or 16	**2** $t = 0$ or 20
3 $b = 49$ or 0	**4** $x = -4$ or -4 (repeating root)
5 $x = -2$ or -3	**6** $m = -7$ or -7 (repeating root)

7 $y = +$ or $- 5$ **8** $t = +$ or $- 10$
9 $g = 12$ or -3 **10** $h = 16$ or -4

Quadratic equations 2 (p 65)

1 $x = 0.16$ or -6.16 **2** $x = -1.5$ or 1.5
3 $x = -0.27$ or -3.73 **4** $x = 0.23$ or -15.23
5 You should have found that you need to find the square root of a negative number. This means the roots of the equation are unreal or imaginary roots.

More on simultaneous equations (p 67)

1 $x = 4, y = 3$ or $x = 3, y = 4$
2 $x = -2.5, y = -2$ or $x = 1, y = 5$
3 $x = 2, y = 2$ or $x = 0, y = 0$
4 $x = 4.93, y = 2.07$
5 $x = 3, y = 4$ or $x = -\frac{1}{2}, y = 7\frac{1}{2}$

Congruence and similarity (p 69)

1 RHS congruence test works.
2 12 **3** (a) 9.2 cm (b) 11.5 cm
4 Neither. Congruence tests do not apply.

Trigonometry in any triangle (p 71)

1 $1667 m^2$ **2** 6.78^0

The sine rule (p 73)

1 $c = 10.06$ cm **2** $WY = 13.12$ cm
3 $x = 16.02$ cm **4** $y = 10.55$ cm
5 $c = 38.77°$ or $142.23°$ **6** $z = 34.25°$ or $145.75°$

The cosine rule (p 75)

1 (a) $44.42°$ (b) $18.57°$
   (c) $95.22°$ (d) $94.41°$
2 (a) 14 (b) 10.44
3 118.21m **4** 177.4 km

Problems in three dimensions (p 77)

1 (a) $\sqrt{29} = 5.4$ cm to 1 d.p.
   (b) $\sqrt{525} = 22.9$ cm to 1 d.p.
   (c) $\sqrt{3} = 1.7$ units to 1 d.p.
2 The length of each slanted edge is $\sqrt{450} = 21.2$ cm to 1 d.p. The total length is therefore $40 + 4\sqrt{450} = 124.9$ cm to 1 d.p. or 1.249 m.
3 $AC = \dfrac{10}{\tan 16°}$ $BC = \dfrac{10}{\tan 32°}$

$\tan \angle CAB = \dfrac{BC}{AC} = \dfrac{10}{\tan 32°} \div \dfrac{10}{\tan 16°} = \dfrac{\tan 16°}{\tan 32°}$

So $\angle CAB = \tan^{-1} \dfrac{\tan 16°}{\tan 32°} = 24.6°$ to 1 d.p. So the bearing of B from A is 025°, to the nearest degree.

Transformations (p 79)

Enlargements (p 81)

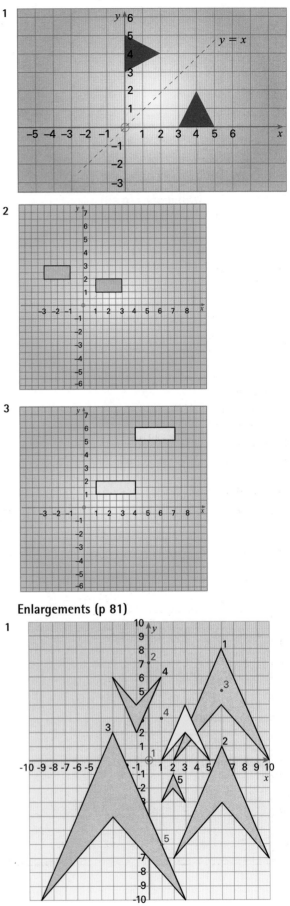

Functions (p 83)

1 (a) 10 (b) 17 (c) 26
   (d) 1 (e) 2 (f) $1\frac{1}{9}$

2 (a) $2x^2 + 4x + 2$ (b) $32x^2$
   (c) $2x^2 + 12x + 18$ (d) $8x^2$
   (e) $\frac{x^2}{2}$ (f) $2a^2x^2 + 4axb + 2b^2$

3 (a) $f^{-1}(x) = \sqrt{x + 3}$ (b) $g^{-1}(x) = x^3 - 2$
   (c) $h^{-1}(x) = \sin^{-1}(1 - x)$ (d) $k^{-1}(x) = \dfrac{x - b}{a}$

Translating graphs (p 85)

1 (a) $y = x^2 + 4$ (b) $y = (x + 3)^2$
   (c) $y = x^2 - 5$ (d) $y = (x - 7)^2$
   (e) $y = (x + 2)^2 + 6$, or $y = x^2 + 4x + 10$
   (f) $y = (x - 4)^2 - 1$, or $y = x^2 - 8x + 15$

2 $3x + 7 = 3(x + 2) + 1$, so

   $$\frac{3x + 7}{x + 2} = \frac{3(x + 2) + 1}{x + 2} = 3 + \frac{1}{x + 2} \text{ or } \frac{1}{x + 2} + 3.$$

   So from $y = \frac{1}{x}$, x has been replaced by $x + 2$ and 3 has been added. Therefore, $y = \frac{1}{x}$ has been translated 2 units to the left and 3 up: along vector.

3 (a) $x^2 + 8x + 12 = x^2 + 8x + 16 - 4 = (x + 4)^2 - 4$.
   So the translation is along vector $\begin{pmatrix} -4 \\ -4 \end{pmatrix}$.
   (b) $(-4, -4)$
   (c) Your sketch should show the curve $y = x^2$, translated 4 units left and 4 units down.

4 The graph is $y = \tan x$, translated with the vector $\begin{pmatrix} 90° \\ 2 \end{pmatrix}$.
   So the equation is $y = \tan(x - 90°) + 2$.

Stretching graphs (p 87)

1 For each part, write down the equation of the object graph.

	object graph	image graph
(a)	$y = x^3$	$y = 3x^3$
(b)	$y = x^2$	$y = \frac{1}{2}x^2 = \frac{x^2}{2}$
(c)	$y = x^2$	$y = (\frac{1}{2}x)^2 = \frac{x^2}{4}$
(d)	$y = (x + 1)^2$	$y = (10x + 1)^2$
(e)	$y = x^3 + 5x^2$	$y = (\frac{x}{5})^3 + 5(\frac{x}{5})^2 = \frac{x^3}{125} + \frac{x^2}{5}$
(f)	$y = \frac{1}{x}$	$y = \frac{1.5}{x} = \frac{3}{2x}$

2 For each part, describe the transformation given by the change in equation.

	object graph	transformation
(a)	$y = x^2$	5 × vertical stretch
(b)	$y = x^2$	3 × horizontal stretch
(c)	$y = x^2 + 2x$	4 × vertical stretch
(d)	$y = x^2 + 2x$	squash to half size horizontally
(e)	$y = \tan x$	2 × horizontal stretch
(f)	$y = \frac{1}{x}$	10 × vertical stretch or 10 × horizontal stretch!

Vectors 1 (p 89)

1 $\begin{pmatrix} 4 \\ 5 \end{pmatrix}$ 2 $p + q = \begin{pmatrix} 4 \\ 3 \end{pmatrix}$, $p - q = \begin{pmatrix} 0 \\ -1 \end{pmatrix}$

Vectors 2 (p 91)

1 (a) $a + b = \begin{pmatrix} 8 \\ 5 \end{pmatrix}$ magnitude = 9.43

   (b) $4c = \begin{pmatrix} -8 \\ 8 \end{pmatrix}$ magnitude = 11.31

   (c) $2d + b - d = d + b = \begin{pmatrix} 7 \\ -3 \end{pmatrix}$ magnitude = 7.62

   (d) $na + ma = (n + m)a = \begin{pmatrix} 1 \\ 5 \end{pmatrix}$ magnitude = 5.10

   (e) $a + 3d - c - 2a = 3d - c - a = \begin{pmatrix} 1 \\ -16 \end{pmatrix}$
   magnitude = 16.03

2 (a) $\begin{pmatrix} 2 \\ 5 \end{pmatrix}$ (b) $\begin{pmatrix} 6 \\ -1 \end{pmatrix}$ (c) $\begin{pmatrix} 4 \\ -6 \end{pmatrix}$
   (d) $\begin{pmatrix} -4 \\ 6 \end{pmatrix}$ (e) $\begin{pmatrix} 4 \\ 2 \end{pmatrix}$

Motion graphs (p 93)

1 (a) Check the coordinates of your points. The curve should be parabolic (the equation is $h = 0.9t^2$).
   (b) The correct value for the velocity is 9 m/s.

2 (a)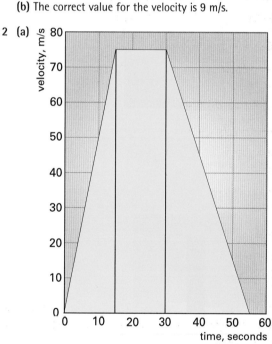

   (b) 2625 m

Circle geometry (p 95)

1 $a = 61°$ 2 $x = 60°$, $2x = 120°$
3 $x = 23°$, $y = 67°$ 4 $x = 24°$, $y = 66°$, $z = 67°$

Circles: arcs and sectors (p 97)

1 (a) 3.49 cm (b) 2.51 m (c) 161 cm
2 (a) 154 cm² (b) 179 mm² (c) 1750 km²
3 Arc length = 17.3 cm to 1 d.p.: perimeter = 39.3 cm to 1 d.p.
4 Radius of outer circle = 50 cm.
   Sector angle = 225°

Area of outer sector = 4908.73... cm_.
Area of inner sector = 1227.18... cm_.
Area of shape = 3681.55... cm_ = 0.37 m² to 2 d.p.

Volume and surface area (p 99)

1 500cm³
2 1232π = 3870 cm², to the nearest cm²
3 4500π = 14 137 cm³, to the nearest cm³
4 324π = 1018 cm³, to the nearest cm³
5 144π = 452 mm², to the nearest mm²
6 Perimeter of triangle = 12 + 5 + 13 = 30 cm
   Area of triangle = 30 cm²
   Total surface area = 2 × 30 + 30 × 20 = 660 cm²

Cumulative frequency (p 101)

1

Age	No of people (Millions)	Cumulative freq
under 10	16	16
10–19	12	28
20–29	17	45
30–39	16	61
40–49	15	76
49–69	10	86
69–89	4	90

2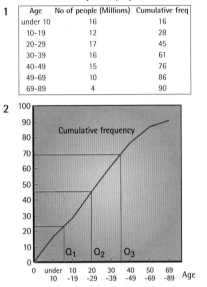

3 Median = Q_2 = 29
   Interquartile range = $Q_3 - Q_1$ = 34 − 16 = 18 (approx.)

Histograms (p 103)

1 (a)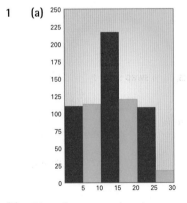

(b) 32 students were absent.

2 (a) and (b)

Height (h cm)	Frequency density	Frequency
$0 \leqslant h < 20$	0.15	3
$20 \leqslant h < 30$	2.2	22
$30 \leqslant h < 40$	2.7	27
$40 \leqslant h < 50$	3.1	31
$50 \leqslant h < 55$	2.8	14

(c) Graham has 97 plants.

Dispersion (p 105)

1 (a) 2.95 (2dp)   (b) 1.67 (2dp)   (c) 2.36 (2dp)
2 (a) mean = 38.4, standard deviation = 3.68 (2dp)
   (b) mean = 18, standard deviation = 4.76 (2dp)
3 (a) mean = 73.2   (b) standard deviation = 13.16
   (c) mean = 78.2 and standard deviation 13.16

Probability: the OR rule (p 107)

1 0.15
2 (a) $\frac{5}{6}$ (b) $\frac{1}{2}$
3 (a) 0.13   (b) 0.7
4 (a) 0
   (b) 1. If I want to get to work, I will certainly have to use one of these modes of transport.
5 (a) $\frac{15}{26}$ (b) $\frac{18}{26}$

Probability: the AND rule (p 109)

1 (a) $\frac{1}{2}$ (b) $\frac{1}{2}$ (c) $\frac{1}{4}$
2 (a) $\frac{1}{16}$ (b) $\frac{1}{169}$ (c) $\frac{100}{169}$
3 (a) $\frac{25}{81}$ (b) $\frac{40}{81}$

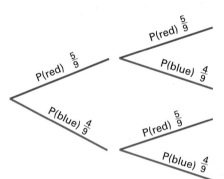

Answers to check questions

Fractions and decimals

The lowest ten are:

$$\overline{2},\ \overline{4},\ \overline{5},\ \overline{8},\ \overline{10},\ \overline{16},\ \overline{20},\ \overline{25},\ \overline{32},\ \overline{40}$$

Zeroes at the left-hand end of an integer are irrelevant.
Because the decimal 'goes on forever'.

Working with indices

2^9 (512)

7^5 (16807)

$(4^3)^3 = 4^9$

$\frac{1}{32}$

$4\sqrt{x}$

Numbers in standard form

7.05×10^2 9.7×10^8

For example, squaring 10^{50}.
Square (when on a single key), square root, trig functions, reciprocal, etc.

Ratio, proportion, percentages

Any number (e.g. 2 : 3 : 6 : 1)
At £1 = €0.63, £50 = €80.65
0.85, 1.3, 0.6

Surds

The square root of the number written using a $\sqrt{}$, not an approximate decimal.
Looking for a factor inside the surd which is a perfect square.
Surds of the same type may be added together.

$\frac{3}{\sqrt{10}}$ = .95 to two d.p.

.24 to two d.p.

Solving simple equations

The number multiplying x.

$y = 4;\ k = \frac{2}{9}$

$n = -1\frac{1}{3};\ r = \frac{1}{2}$

Rearranging forumale

$x = p - y$

$\frac{p}{6} - \frac{q}{6}$

It changes its sign.

$n = \frac{W}{t - s}$

Inequalities

x lies between -3 and 6 and is allowed to be = to 6.
To turn $4x$ into x.

Lines and equations

On this line, as x increases, y decreases.
If the line crosses the y-axis, you can just read it off.
$ax + by = k$ can be rearranged to
$y = -\frac{a}{b}x + \frac{k}{b}$, so $m = -\frac{a}{b}$ and $c = \frac{k}{b}$

-2

Simultaneous equations and graphs

When they are parallel.
No.

Simultaneous equations and algebra

$x = 8, y = -1$
Multiply equation (i) by 5 and multiply equation (ii) by 2.

Using brackets in algebra

$15x^3$ $5xy$ $x = 6$

$5 \times (-6) = -30$

Multiplying bracketed expressions

$2wy + 2wz + xy + xz$

$x \times (-x) = -x^2$

$x^2 + 10x + 25$

$x^2 - 49$

Regions

$x \leqslant 4$

$2 \leqslant x \leqslant 4$ and $-1 \leqslant y \leqslant 5$

Because $y = 1$ is included in the inequality.

Sequences and formulae

$u_3 = 2 \times 3 + 1 = 7$

$u_{100} = 5 \times 100 + 4 = 504$

$u_n = 5n - 1$

More on formulae

$x = 2.86$ to 2 d.p.
Subtract 5 from both sides, then swap sides so that t is on the left-hand side.

Equations of proportionality

$x = 2400$
The general equation is $xy = k$: the one in this example is $xy = 18$.
$y = 3.125x^3$ and when $x = 2, y = 25$.

Quadratic functions

The y-axis.

Cubic functions

$3x$ means $3 \times x$.
x^3 means $x \times x \times x$.
The same as $y = x^3$, translated 4 units in the y direction.

Graphing reciprocal functions

You always get 1.

That term disappears from the equation.

Solving equations with graphs

$x = 1\frac{1}{2}$

Trial and improvement

Because you cannot be sure how much more than 1.73 it is and it may affect the rounding.

The previous values still give answers that could round to either 1.73 or 1.74.

More complicated indices

$x = x^1$ and $y = y^1$

m^{12}

$y = y^1$: using the rules of indices, $3 - 1 = 2$.

The answer is always 1.

Indices and equations

To divide by a power, subtract indices. Subtracting (-4) is the same as adding 4, which represents mulitplying x^4.

$\sqrt[4]{x}$ is the same as as $x^{\frac{1}{4}}$

$(\sqrt[4]{x})^3 = (x^{\frac{1}{4}})^3 = x^{\frac{3}{4}}$

Factorising quadractic expressions

$(x + 4)(x - 6)$

6 is the HCF that was extracted earlier.

They are easier to pick out when one set of brackets is nested inside another.

The difference of two squares

$(x - 2)(x + 2)$

$= x(x + 2) - 2(x + 2)$

$= x^2 + 2x - 2x - 4$

$= x^2 - 4$

It's the square root of $81y^2$.

$3(4p - 5q)(4p + 5q)$

$= (12p - 15q)(4p + 5q)$

$= 12p(4p + 5q) - 15q(4p + 5q)$

$= 48p^2 + 60pq - 60pq - 75q^2$

$= 48p^2 - 75q^2$

Algebraic fractions

A number or term that divides exactly into two or more terms.

It was cancelled in step 3.

Writing something as a fraction over 1 doesn't change its value.

$\frac{6x - 3}{4}$

Quadratic equations 1

$x = 0$ or $x = -5$

$6x - 5 = 0$ means $6x = 5$ and so $x = \frac{5}{6}$.

$x + 1 = 0$ means $x = -1$.

$144 = 12^2$

$169 = 13^2$

Quadractic equations 2

Quadratic equations have two solutions.

$ac = 16$ and $b = 8$; the required factors are 4 and 4.

More on simultaneous equations

By substituting for x into equation (ii).

Equation (ii) says that $x = y$.

Equation (i) is a curve, (ii) is a straight line. The line is a tangent of the curve.

Congruence and similarity

The size of the shape is being increased.

9 cm and 6 cm are corresponding sides.

Trigonometry in any triangle

$180° - 43.81° = 136.19°$

The sine rule

$a = \dfrac{b \sin A}{\sin B} = \dfrac{c \sin A}{\sin C}$

The cosine rule

$\cos A = 0$ and therefore $2bc \cos A = 0$. So $a^2 = b^2 + c^2$, which is Pythagoras' rule. a is the hypotenuse of the right-angled triangle.

By using the inverse cosine, $\cos^{-1}$.

Problems in three dimensions

There are four. They are all equal because each one has to 'cover' the length, width and height of the cuboid.

From the angle sum of $\triangle AHC$

Transformations

The reflection raises its 'right' hand.

$\begin{pmatrix} -5 \\ 1 \end{pmatrix}$

Enlargements

If the triangle is X"Y'"Z", then X'X" = AX', etc., or AX" = 2AX, etc.

A ➡ $(\frac{1}{3}, 1\frac{1}{3})$

B ➡ $(1, 1\frac{1}{3})$

C ➡ $(1, \frac{1}{3})$

The vertices of the enlargement are at (-6, -2), (-6, -8) and (-2, -8).

Functions

$f(2m) = (2m)^3 = 8m^3$

51

$h^{-1}(x) = \frac{1}{x}$. Some functions are 'self-inverse'.

Translating graphs

The graph of $y = \cos x$ is translated 2 units 'down'.

The graph of $y = x^2$ is translated 1 unit 'to the right'.

For example, (2, 4) on $y = x^2$ translates to (5, 2) using the vector.

Substituting $x = 5$ into $x^2 - 6x + 7$ gives $25 - 30 + 7 = 2$.

Stretching graphs

The graph is a 2 × vertical stretch of $y = \sin x$.

The graph is $y = x^2$ squashed to half size horizontally.

Vectors 2

$\frac{1}{2}v = \begin{pmatrix} \frac{3}{2} \\ 2 \end{pmatrix}$

$4p - q = \begin{pmatrix} -12 \\ -4 \end{pmatrix}$

$\overrightarrow{CD} \begin{pmatrix} 5 \\ -12 \end{pmatrix}$, midpoint = (4.5, 2)

$\overrightarrow{MN} = \sqrt{-2^2 + 8^2} = \sqrt{68} = 8.25$

Motion graphs

0.8 m/s²

−1 m/s²

13.75 m/s

Circles: arcs and sectors

$\frac{1}{8}$

46.2 cm to 3 s.f.

$37\frac{1}{2}\pi$ or $\frac{75\pi}{2}$

Volume and surface area

It's a prism with a rectangular cross-section.

V = 2413 cm³ to nearest cm³. S = 1056 cm² to nearest cm².

546.10 cm² to d.p.

Cumulative frequency

It contains the middle 50% of the data.

You find the 25%, 50% and 75% marks on the cumulative frequency axis and work across to the age axis.

Histograms

Frequency = frequency density × class width.

Dispersion

The mean of the squares minus the square of the mean.

Probability: the OR rule

$1 - \frac{1}{13} = \frac{12}{13}$

They are both the same.

Probability: the AND rule

Yes. The result of the first flip does not affect the second.

$\frac{13}{28}$

$\frac{76}{231}$

Glossary

A

Acute angle An angle less than 90° (less than a right angle).

Algebra The branch of mathematics that deals with the general case. Algebra involves the use of letters to represent variables and is a very powerful tool in problem-solving.

Angle A measure of space between two intersecting lines.

Arc Part of the circumference of a circle.

Area A measure of two-dimensional space.

Average See mean, median and mode.

B

Bar chart A frequency chart, where the frequency of the data is proportional to the height of the bar.

Bearing A measure of an angle used in navigation.

C

Chord A line across a circle, not passing through the centre.

Circumference The complete boundary of a circle.

Coefficient The number in front of the variable that shows the magnitude (size) of the variable, for example in $2x$ the coefficient of x is 2 and in $5y$ the coefficient of y is 5.

Compound interest The type of interest that is paid by most banks, where the interest is added to the principle invested and then in subsequent years interest is earned by the original interest.

Correlation The connection between two variables. It can be a positive correlation; in other words, as one of the variables increases, the other decreases in the same proportion. Alternatively, it could be a negative correlation, where as one of the variables increases, the other decreases in the same proportion. An example is smoking. There is a definite correlation between smoking and lung cancer and other diseases. You may be unfortunate and get lung cancer without ever smoking but the vast majority of lung cancer sufferers are also smokers. The more you smoke, the more you increase your chances of becoming ill.

Cube A three-dimensional solid which has sides of equal length.

Cube root The cube root of a number, for example 64, is the number that you need to multiply by itself and then by itself again to make, in this case, 64. Mathematically, this must be 4, because $4 \times 4 = 16$ and then $16 \times 4 = 64$. We write this as $\sqrt[3]{64}$

and this reads as 'the cube root of 64'.

Cumulative frequency Think of this as a running total graph. Add up the frequencies as you go through the data. The cumulative frequency is drawn as a typical 'S' shaped graph. The curve is also known as an ogive.

Cylinder A three-dimensional solid with a uniform cross-section that is a circle.

D

Denominator The lower part of a fraction.

Diameter The distance across a circle from one point on the circumference to another point on the circumference, passing through the centre.

Dimension A dimension is a length. Formulae can be analysed to determine what type of formulae they may be. For example, $A = lb$, where l and b are both lengths, could be an area. This is because a formula for area must contain two lengths. A formula for volume must contain three lengths.

Distribution A set of data.

Dodecahedron A regular solid with 12 faces.

E

Equation A mathematical statement, usually in algebra, where two sides of the statement are equal. The aim is to work out the value of the unknown by manipulating the equation.

Equation, linear An equation where the highest power of any of the terms is 1.

Equation, quadratic An equation where the highest power of any of the terms is 2.

Equilateral triangle A triangle that has all three sides of equal length and all three angles equal 60°.

Estimate The process of comparing the size of a property of an object with a known quantity.

Euro Europe-wide currency that was established in 1999 and introduced into many European countries in 2002.

F

Factor A number or variable that divides into other numbers or variables without a remainder.

Factorisation The process of extracting the highest common factors from an expression.

Formula An algebraic statement that is the result of previously established work. A formula is accepted to be correct and can be used in subsequent work. For example, $A = \pi r^2$ is accepted as the formula for the area of a circle. It is not necessary to prove this formula every time we need to use it.

Frequency polygon A way of displaying grouped data where the mid-values of the class intervals are joined by lines.

Function A rule which applies to one set of quantities and how they relate to another set.

G

Generalising A process in mathematical thinking, where a general rule, usually expressed in algebra, is determined.

Gradient The measurement of the steepness of a line. It is the ratio of the vertical to the horizontal distance. Where a line slopes from bottom left to top right, the gradient is positive. Where the line slopes from top left to bottom right, the gradient is negative.

Graph A visual display, used to show information from data distributions, to create a visual understanding of the nature of the distribution.

Graphical calculator An electronic calculator on which it is possible to draw graphs.

H

Highest common factor The highest factor that will divide exactly into two or more numbers.

Hypotenuse The longest side of a right-angled triangle.

I

Identity An algebraic expression where the two sides of the identity are equal because they are not different, for example
$2x = x + x$.

Imperial units The traditional units of measure which were once used in the UK and the USA, such as feet, inches, pints and gallons.

Improper fraction A fraction where the numerator is of a higher value than the denominator.

Independent event An event whose outcome is not dependent on the outcomes of other events.

Inequality An algebraic statement which is unequal.

Integer A whole number.

Interquartile range The spread of the middle 50% of a distribution.

Irrational number A number that cannot be totally determined, for example $\sqrt{2}$.

Isosceles triangle A triangle with two equal sides and two equal angles, and where the equal sides are opposite the equal angles.

L

Line of symmetry A line that bisects a shape so that each part of the shape reflects the other.

Lowest common multiple The lowest number that two or more numbers will go into, for example the lowest common multiple of 4 and 8 is 8, because 8 is the lowest number that both 4 and 8 will go into.

M

Mean The arithmetic average, which is calculated by adding all of the items of data and then by dividing the answer by the number of items of data.

Median The second type of average. It is the middle value in a set of data which are in order of size from highest to lowest or lowest to highest.

Metric The system of measurements usually used in Europe and the most commonly used system in science.

Mode The value that occurs the most often in a set of data.

N

Negative number A number with a value of less than 1.

Net The pattern made by a three-dimensional shape when it is cut into its construction template and then laid flat.

Number, cube The sequence of cube numbers is 1, 8, 27, 64, 125, etc. It is made up from 1 x 1 x 1, 2 x 2 x 2, 3 x 3 x 3 and so on.

Number, prime A number with two and only two factors. The factors are the number itself and 1.

Number, square The sequence of square numbers is 1, 4, 9, 16, 25, 36, etc. It is made up from 1 x 1, 2 x 2, 3 x 3 and so on.

Numerator The upper part of a fraction.

O

Obtuse Refers to any angle greater than 90° but less than 180°.

P

Parallel lines Lines that are equidistant along the whole of their length, in other words they are a constant distance apart along the whole of their length.

Perimeter The total distance around the boundary of a shape.

Perpendicular lines Lines that meet at right angles.

Pie chart A chart in the shape of a circle, where the size of the sector shows the frequency.

Polygon A many-sided shape.

Powers The power to which a number is raised, also called 'indices'.

Prism A solid with a uniform cross-section.

Probability The study of the chance of events occurring.

Pythagoras' rule (theorem) The statement of a relationship between the three sides of a triangle, which is 'the square of the hypotenuse of a right-angled triangle is equal to the sum of the squares of the other two sides'. Using algebra, it is written as $h^2 = a^2 + b^2$.

Q

Quadratic equation An equation where the highest power is 2.

Quadrilateral A plane four-sided shape.

R

Radius The distance from the centre of a circle to any point on the remaining part of the circumference. So the radius is half of the diameter of the same circle.

Ratio The relationship between one quantity and another.

Rational number A number that can be written as a fraction.

Reflex angle Any angle greater than 180°.

Right angle Any angle that is equal to 90°.

S

Scalene triangle A triangle where all of the sides and all of the angles are of different sizes.

Sequences A set of numbers having a common property. A mathematician can work out what the numbers have in common and so predict further numbers in the sequence.

Standard deviation The square root of the variance. It is a measure of dispersion.

Standard form A way of writing very large or very small numbers, using powers of 10.

Surds Numbers often expressed as the square roots of a number.

T

Tangent A straight line that touches a circle at one point on its circumference.

V

Variance The mean of the deviations from the mean of a set of data.

VAT Value Added Tax.

Last-minute learner

Number

Types of number
- **Integers** are positive and negative whole numbers. Any integer can be written as a product of **prime factors**, eg. $24 = 2^3 \times 3$, $450 = 2 \times 3^2 \times 5^2$.
- **Rational numbers** include integers and fractions. Numbers like $\sqrt{2}$, $\sqrt[3]{10}$ and π are **irrational**.

Indices
- **Index laws:** in algebra, $a^n \times a^m = a^{(n+m)}$, $a^n \div a^m = a^{(n-m)}$.
- A negative power is just the **reciprocal** of the positive power. $2^{-2} = \frac{1}{2^2} = \frac{1}{4}$, $2^{-3} = \frac{1}{2^3} = \frac{1}{8}$, etc.
- **Fractional** indices mean **roots**.
- Numbers in **standard index form** consist of a power of 10 multiplying the significant digits of the number.
  2 million $= 2 \times 1\,000\,000 = 2 \times 10^6$.
  $2\,500\,000 = 2.5$ million $= 2.5 \times 10^6$.
- Numbers less than 1 need a negative index, e.g. $0.002\,13 = 2.13 \times 10^{-3}$.
- You may be required to leave some answers in **surd** form. Use facts such as $\sqrt{a} \times \sqrt{b} = \sqrt{ab}$.
- Rationalise denominators of types $\sqrt{a}$ and $(a + \sqrt{b})$ by multiplying top and bottom by $\sqrt{a}$ and $(a - \sqrt{b})$, respectively.

Fractions and decimals
- To convert a fraction to its equivalent decimal, divide numerator by denominator.
- To convert a terminating decimal to a fraction, use a sufficiently large power of 10 as denominator. For recurring decimals, multiply by a power of 10 determined by the period of the decimal, then subtract to eliminate the recurring part.

Proportional quantities
- If y is **directly proportional** to x, this is written $y \propto x$. That means $y = kx$, where k is the **constant of proportionality**. Two proportional amounts plotted against each other on a **graph** give a straight line through the origin. If y is proportional to a power of x, this is written $y \propto x^n$, meaning $y = kx^n$.
- If y is **inversely proportional** to a power of x, this is written $y \propto \frac{1}{x^n}$, meaning $y = \frac{k}{x^n}$.

Algebra

Formulae and expressions
- **Substitution** is replacing letters in a formula, equation or expression by numbers (their **values**). Be careful to **evaluate** parts of the formula in the correct order.
- **Expressions** in algebra are made up of a number of terms added or subtracted together. Each **term** is made up of letters and numbers multiplied or divided together. Combine **like terms** to simplify an expression.
- A **formula** usually has its **subject** on the left-hand side of the equals sign and an expression on the right-hand side. Any letter in a formula can become the subject by **rearranging** it. As long as you do the same thing to both sides of your formula, it is still true.

Multiplying and dividing terms
- When **multiplying two terms** together, multiply the numbers first, then multiply the letters in turn, using the index rules.
- When **dividing terms**, write the question in fraction style if it's not already written that way and cancel the numbers as if you were cancelling a fraction to lowest terms, then divide the letters in turn.

Expanding brackets
- When a number or letter multiplies a bracket, **everything** inside the bracket is multiplied. Removing the brackets is called **expanding** them.

Factorisation
- **Factorising** is the opposite of expansion. To **factorise**, look for **common factors** between the terms. This process is called **extracting factors**. Sometimes you need to do this in more than one step.
- **Quadratic** expressions can sometimes be factorised into two brackets. First, write down a list of the numbers that could be part of the x^2 term. Write down a list of the numbers that could be part of the number term. Test combinations of these numbers to see if you can match the x term in the expression you want to factorise.
- **Difference of two squares:** $a^2 - b^2 = (a - b)(a + b)$.

Quadratic equations
- Quadratic equations of the form $ax^2 + bx + c = 0$ can be solved by factorisation, or by using the quadratic formula: $x = \frac{-b \pm \sqrt{b^2 - 4ac}}{2a}$

Trial and improvement
- Sometimes you can't find an exact solution to an equation but can find a reasonable **approximation** using **trial and improvement** or a **decimal search**. You use the results of a 'guess' to make better guesses.

Simultaneous equations
- **Simultaneous equations** are pairs of equations with two unknown letters that are both true at the same time. The technique of **elimination** involves adding or subtracting

Last-minute learner

Simultaneous equations cont.

the equations so that one of the letters disappears (is **eliminated**). Sometimes, you have to multiply one or both of the equations. This is to match coefficients in order to add or subtract and eliminate.

- Another method of solution is to express one variable in terms of the other, then **substitute** the resulting expression into the other equation. This is useful when one equation is quadratic.
- Each equation in a simultaneous pair has its own **graph**. The x and y co-ordinates of the point where the graphs **intersect** give the solution.

Inequalities

- Ranges of numbers are described using **inequalities**.
- There are four inequality symbols:

$>$ greater than	$\geqslant$ greater than or equal to
$<$ less than	$\leqslant$ less than or equal to

'All the numbers that are 3 or less' is described by the inequality $x \leqslant 3$ (or $3 \geqslant x$).
- Sometimes inequalities can be combined. Suppose that $x < 2$ and $x \geqslant -3$. This makes a **range inequality**, $-3 \leqslant x < 2$.

- Solve inequalities using the techniques for equations, e.g. the solution of $2x - 1 > 5$ is $x > 3$.
- A line (e.g. $y = 5 - x$) divides a co-ordinate grid into two **regions**. The region above the line is $y > 5 - x$ and the one below is $y < 5 - x$. The line can be included in the region by using $\geqslant$ or $\leqslant$.

Sequences and functions

- **Sequences** are made up of a succession of **terms**. Each term has a **position** in the sequence: 1st, 2nd, etc. A **linear** sequence has the formula $u_n = an + b$.
- **Quadratic** sequences have formulae of the form $u_n = an^2 + bn + c$. To analyse quadratic sequences, look at the **second** difference row.
- A **function** is a rule applied to a variable, e.g. if $f(x) = x^2 - 3$, then $f(5) = 22$ and $f(x + 1) = (x + 1)^2 - 3 = x^2 + 2x - 2$.

Transformations of graphs

The graph of $y = f(x)$ is transformed as follows:
$y = f(x) + a$: translation a units in the y direction.
$y = f(x + a)$: translation $-a$ units in the x direction.
$y = af(x)$: stretch in the y direction, scale factor a.
$y = f(ax)$: stretch in the x direction, scale factor $\frac{1}{a}$.

Accuracy of measurements

- Suppose the length L of a piece of wire is 67 mm, to the nearest mm: then $66.5 \leq L < 67.5$. If L is given to the nearest 0.1 mm (1 d.p.) instead, then $66.95 \leq L < 67.05$.

Speed, distance and time

- Units of speed are metres per second (m/s), kilometres per hour (km/h), miles per hour (mph), etc.
- Use the 'd-s-t triangle'. Cover up the letter you want to work out: the triangle gives the formula.

- On a distance-time graph, the gradient of the tangent gives the speed at a given time.
- On a velocity-time graph, the gradient of the tangent gives the acceleration at a given time, and the area bounded by the graph and the time axis gives the distance travelled.

Mensuration – length and distance

- In any right-angled triangle, the **hypotenuse** is the side opposite the right angle.

- Pythagoras' rule: in any right-angled triangle, $h^2 = a^2 + b^2$.

Use it to calculate the diagonal of a rectangle, or the distance between two points on a co-ordinate grid.
- The **space diagonal** of a cuboid with dimensions a, b and c is $d = \sqrt{a^2 + b^2 + c^2}$
- The **circumference** of a circle of diameter d (radius r) is $C = \pi d = 2\pi r$. $\pi \approx 3.142$. The length of an arc subtending $x°$ at the centre is $\frac{x}{360}$ of the circumference.
- When a wheel turns once (makes one **revolution**), the distance moved by whatever it's attached to (e.g. a car, bike, etc.) is the same as the circumference of the wheel.

Trigonometry

- In a right-angled triangle, $\sin x = \dfrac{\text{opposite side}}{\text{hypotenuse}}$.

  $\cos x = \dfrac{\text{adjacent side}}{\text{hypotenuse}}$, $\tan x = \dfrac{\text{opposite side}}{\text{adjacent side}}$
- To find an angle in a right-angled triangle with known sides, calculate the trig ratio, then use the **inverse trig function** ($\sin^{-1}$, etc.).
- In any triangle ABC with sides a, b and c opposite their corresponding vertices, you can use:
- the sine rule: $\dfrac{a}{\sin A} = \dfrac{b}{\sin B} = \dfrac{c}{\sin C}$;

- the cosine rule: $a^2 = b^2 + c^2 - 2bc \cos A$ (also $b^2 = c^2 + a^2 - 2ca \cos B$ and $c^2 = a^2 + b^2 - 2ab \cos C$).

Last minute-learner

Area

- Using l for length and w for width, the area A of a rectangle is given by $A = lw$.
- Triangles, parallelograms and trapezia all share an important measurement: the **perpendicular height**. These are the area formulae:

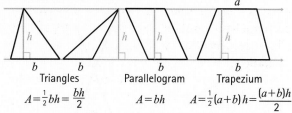

Triangles	Parallelogram	Trapezium
$A = \frac{1}{2}bh = \frac{bh}{2}$	$A = bh$	$A = \frac{1}{2}(a+b)h = \frac{(a+b)h}{2}$

- The area of the circle of radius r is $A = \pi r^2$. The area of a sector subtending $x°$ at the centre is $\dfrac{x}{360}$ of the total area.
- In any triangle ABC as described above, the area is given by $A = \frac{1}{2}ab \sin C$.

Surface area

- The surface area of a solid object is the combined area of all the faces on the outside. Curved surfaces on spheres, cones and cylinders form part of the surface area too.
- For a prism with cross-section of perimeter P and area A, the total surface area $S = 2A + Pl$. For a cylinder, $S = 2\pi r^2 + 2\pi rl = 2\pi r(r + l)$.
- Cones need an extra measurement, the slant height s. The curved surface is πrs, so the total surface area is $\pi r^2 + \pi rs = \pi r(r + s)$.
- The surface area of a sphere is $4\pi r^2$.

Volume

- There are four basic volume formulae.

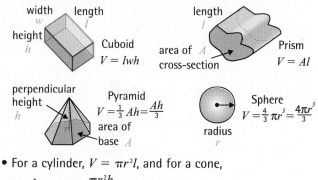

Cuboid
$V = lwh$

area of cross-section

Prism
$V = Al$

Pyramid
$V = \frac{1}{3}Ah = \frac{Ah}{3}$
area of base A

Sphere
$V = \frac{4}{3}\pi r^3 = \frac{4\pi r^3}{3}$
radius r

- For a cylinder, $V = \pi r^2 l$, and for a cone,
$V = \frac{1}{3}\pi r^2 h = \dfrac{\pi r^2 h}{3}$.

$1 \text{ cm}^3 = 1 \text{ ml}$, $1 \text{ l} = 1000 \text{ cm}^3$, $1000 \text{ l} = 1 \text{ m}^3$.

Angles and Shapes

- Whenever lines meet or **intersect**, the angles they make follow certain rules.

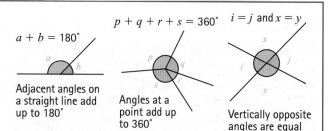

$a + b = 180°$

Adjacent angles on a straight line add up to 180°

$p + q + r + s = 360°$

Angles at a point add up to 360°

$i = j$ and $x = y$

Vertically opposite angles are equal

- Three types of relationship between angles are produced when a line called a **transversal** crosses a pair of **parallel** lines.

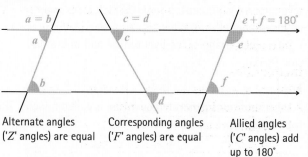

$a = b$

Alternate angles ('Z' angles) are equal

$c = d$

Corresponding angles ('F' angles) are equal

$e + f = 180°$

Allied angles ('C' angles) add up to 180°

- The **exterior angles** of a polygon always add up to exactly 360°.
- Every type of polygon has its own **interior angle sum**. You can calculate it using any of these formulae: n is the number of sides and S is the angle sum.
  $S = (n - 2) \times 180°$ $S = (180n - 360)°$
  $S = (2n - 4)$ right angles
- Work out the interior angles for **regular** polygons in two ways: work out the angle sum, then divide by the number of sides; or divide 360° by the number of sides to find one exterior angle, then take this away from 180°.

Transformations

- Mathematical transformations start with an original point or shape (the object) and transform it (into the **image**).
- A **translation** is a 'sliding' movement, described by a **column vector**, eg. $\begin{pmatrix} 5 \\ -4 \end{pmatrix}$.
- In a rotation, specify an angle and **centre of rotation**.

Given a rotation, to find the centre:
- join two pairs of corresponding points on the object and image
- draw the perpendicular bisectors of these lines
- the point of intersection is the centre of rotation.
- Reflection in any line is possible, but the most likely ones you will be asked to use are these:
  - horizontal lines ($y = a$ for some value of a);
  - vertical lines ($x = b$ for some value of b);
  - lines parallel to $y = x$ ($y = x + a$ for some value of a);
  - lines parallel to $y = -x$ ($y = a - x$ for some value of a).

Last-minute learner

Shape, space, and measures cont.

• To describe an enlargement you need to give a **scale factor** and a centre of enlargement. To find the **centre of enlargement**, draw lines through corresponding points on the object and image. These all intersect at the centre of enlargement. Enlargements are mathematically **similar** to their objects.

Congruence tests
If the following features of two triangles match, the triangles are congruent: 3 sides (**SSS**); 2 sides and the included angle (**SAS**); 2 angles and a side (**AAS**); in a right-angled triangle, the hypotenuse and one other side (**RHS**);

Loci
A set of positions generated by a rule is called a **locus**. The four major types are as follows:

A fixed distance from a fixed point: a circle.

A fixed distance from a straight line: two parallel straight lines.

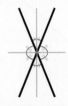

Equidistant from two fixed points: the perpendicular bisector of the points.

Equidistant from two straight lines: the bisectors of the angles between the lines.

Often, you need to combine information from two or more loci. This will lead to a region or area, a line segment, or one or more points.

Data handling

Histograms
• In a **histogram**, **frequency density** (frequency ÷ class width) is plotted against data value. This has the effect that equal frequencies in the data are represented by equal areas on the histogram.

Averages
• The **mode** is the most common value. With grouped data, the group with the highest frequency is called the **modal group** or **class**. The **median** is the middle value in a set, when all the numbers are arranged in order. The **mean** is the sum of the data items, divided by the total frequency.

Dispersion
• The **range** is simply the difference between the smallest and largest data items. The **interquartile range** encloses the middle 50% of a data set. The **variance** is the mean of the squares of the deviations from the mean value: the **standard deviation** is the square root of this.

Probability
• The **OR rule**: when two outcomes A and B of the same event are exclusive, $P(A \text{ or } B) = P(A) + P(B)$.
• **Theoretical probability** is calculated by analysing a situation mathematically.
• **Experimental probability** is determined by analysing the results of a number of trials of the event.
• The **AND rule**: when two events X and Y are independent, $P(X \text{ and } Y) = P(X) \times P(Y)$.
• Use a **tree diagram** to organise multiple events with varying probabilities.